ATLAS
of poetic
BOTANY

Originally published as *Atlas de botanique poétique*.
© Flammarion, Paris, 2016

Design by Karin Doering-Froger

This book was set in DTL Paradox by The MIT Press. Printed and bound
in Spain.

Library of Congress Cataloging-in-Publication Data

Names: Hallé, Francis, author. | Patriarca, Éliane, author. | Butler, Erik,
    1971– translator.
Title: Atlas of poetic botany / Francis Hallé with Éliane Patriarca ;
    translated by Erik Butler.
Description: Cambridge, MA ; London, England : The MIT Press, [2018]
    | Translation of: Atlas de botanique poétique.
Identifiers: LCCN 2018013342 | ISBN 9780262039123 (hardcover : alk.
    paper)
Subjects: LCSH: Botany—Miscellanea. | Plants—Miscellanea.
Classification: LCC QK87 .H35 2018 | DDC 580.1/4—dc23 LC record
    available at https://lccn.loc.gov/2018013342

10 9 8 7 6 5 4 3 2 1

ATLAS
of poetic
BOTANY

FRANCIS HALLÉ

in collaboration with

ÉLIANE PATRIARCA

THE MIT PRESS
CAMBRIDGE, MASSACHUSETTS
LONDON, ENGLAND

CONTENTS

5

THE WORK AT HAND presents a collection of remarkable plants I have observed while working as a botanist in tropical rainforests—or "equatorial forests," as they are also known. My selection features plants that exhibit a strange character, bizarre appearances, and unexpected humor. Their poetry is as vivid as that of bees swarming around a hive, or sea spray adorning the deck of a ship.

As far as the eye can see, equatorial forests harbor an array of immense, if somewhat austere, trees. Although they exude a poetry of their own, on the whole they're more forbidding than entertaining. Perhaps I don't need to invoke these trees to draw attention to the last equatorial forests—a tremendous wellspring of interest and novelty, strangeness and delight—and the need to preserve what little remains of this part of humankind's inheritance. Instead of broaching the tragic topic of deforestation, let me simply defer to Jane Goodall, the great English primatologist, who has devoted her life to studying chimpanzees in the forests of Africa: "Someone once wondered why it is that if a work of man is destroyed, it is called vandalism, but if a work of nature ... is destroyed it is so often called progress" (*Seeds of Hope*, Grand Central Publishing, 2015).

For now, I'll set aside the great trees in favor of plants that are more modest but also more approachable.

This book aims to show that the equatorial forest isn't the "green inferno" that colonialists and adventurers so often confronted. On the contrary, it is a universe of magical allure, and, if only one casts a somewhat sympathetic gaze at the little marvels that present themselves to the visitor at every turn, life here is quite pleasant. There is an abundance of aesthetic satisfaction, wonder, and poetry to be found.

For biology enthusiasts, such an encounter not only makes life more agreeable but yields a veritable firework display of thought-provoking questions. Consider the words of Charles Darwin when, at the age of twenty-two and having just graduated from

Cambridge, he discovered the Brazilian forest: "The day has passed delightfully. Delight itself, however, is a weak term to express the feelings of a naturalist who, for the first time, has wandered by himself in a Brazilian forest. ... Such a day as this brings with it a deeper pleasure than he can ever hope to experience again."

As a botanist, I have tried to use drawings to show to what degree plants—which are not nearly as well known as animals—can prove surprising, enigmatic, and even funny. Although animals represent our initial source of enchantment in these forests, they occupy only a minimal space here. Indeed, the distinction between plants and animals is somewhat obsolete in the equatorial forest, where the most stimulating questions arise from interactions between these two groups of living beings.

I'm often asked: why drawings instead of photographs, which are so easy to use these days? My answer touches on the profound nature of the botanist's work; we stand before vegetal organisms whose main feature, to my eyes, is how profoundly different they are from human beings. It is as if we were visiting a distant planet and encountered a form of extraterrestrial life with which we share no language—a form of life based on principles that are not our own. If we wish to understand this creature, it is best not to rush. That's the first reason to prefer drawings to photos: the quality of time spent is not the same in these two approaches.

To seize an ephemeral moment, as photography does, is to content oneself with limited information. The extended time required for drawing, on the other hand, amounts to a dialogue with the plant: it opens up space for reflection, which is absolutely necessary when one stands face to face with an alien. Drawing represents the work of human thought: the exchange with the object rendered plays a key role, and if questions arise when contemplating the alien, the interview ought to last long enough for an answer to emerge.

A work of human thought: that's another reason to prefer drawing to photography. Both activities represent intellectual work, of course, but in the case of photography, the thinking belongs more to the person who has conceived and constructed the camera than to the person who uses it; the quality of the equipment is what's decisive. In contrast, drawing—which relies more on the individual brain and hand and does so without technological mediation—is more directly the work of the artist.

This explains why botanical drawings so often give rise to a particular emotion, and that's a third reason to prefer them to photography. Contemplating Dürer's *The Great Piece of Turf*, Botticelli's *Primavera*, herbaria by Aldrovandi and Mutis, Redouté's *Les Roses*—or, closer to our own day, Robert Hainard's *Herbier alpin, herbier divin*, Dominique Mansion's pictures in *Flore forestière française*, Adrian D. Bell's *Plant Form*, with drawings by Alan Bryan (Timber Press, 2008), or Philippe Danton's *Monographie de la flore vasculaire de l'archipel Juan Fernandez*—means acknowledging that botanical illustration forms part of a precious tradition that should be respected, renewed, and enriched. Computerized data may yet set our civilization marching backward in great strides.

My hope is that the reader will draw from this book a fresh perspective on the high equatorial forests. It isn't too late to discover what everyone needs: a potent antidote to the woes that attend our modern life in the metropolis.

FRANCIS HALLÉ

RECORDS
AND EXUBERANCE

AN INVASIVE BEAUTY

....

Eichhornia crassipes (Mart.) Solms-Laub.
PONTEDERIACEAE
Water hyacinth

....

THE WATER HYACINTH, with its purplish-blue flower, top petal stained yellow, and dark, feathery roots, is sublime. Bedazzled by this languorous plant floating on the water, the first Europeans to visit the Orinoco Basin brought it back with them, not suspecting that the beauty held a veritable plague in store.

Introduced first to Louisiana in 1884, and then to Mexico, where it quickly became prized as an ornamental plant in garden ponds, the water hyacinth has now spread to every continent. But only in its native environment does it produce seeds.

This aquatic plant displays one of the most spectacular rates of growth on record. Water hyacinths can grow by as much as two to five meters a day. Thanks to its vegetative reproduction, a single plant can clone itself into thirty descendants in twenty-three days, and 1,200 in four months.

Taken out of its natural surroundings, the water hyacinth has no predators (manatees, for instance) and thrives without a problem.

Its clustered leaves cover waterways, lakes, and lagoons with an impenetrable screen. Water hyacinths block the light and capture the oxygen that algae and fish need. They clog up hydro-electric generators and boat propellers, and the evapotranspiration they cause lowers the water level. The plant has colonized the rivers of Africa—the Congo and the Nile—as well as its Great Lakes, Victoria and Tanganyika. China, India, Indonesia, Australia, and Russia have experienced the same scourge on expanses of fresh water.

All efforts to eradicate this invader have failed. The stalks of its leaves, which fill up with air, serve as flotation devices and make it unsinkable.

Taken out of its natural surroundings, the water hyacinth has no predators (manatees, for instance) and thrives without a problem.

....

Crushing them just makes little fragments that, thanks to clonal reproduction, yield a new plant.

This hyacinth's vigor and triumph have prompted a new strategy: now, people try to use the plant instead of fighting it. For instance, it can be dried out and turned into food pellets for livestock; pigs, in particular, are quite fond of it. It is also being used to purify polluted waters, since the plant readily absorbs toxins and heavy metals. Finally, it provides building material for furniture. In Myanmar, Thailand, and Vietnam, its roots are boiled and dried, then twisted into ropes and plaited around a bamboo frame. Such industry keeps the plant from spreading too fast.

The water hyacinth isn't the only ornamental plant that has turned out to be invasive: other offenders include Gabonese tulip trees in Réunion, pampas grass in Spain, mimosas on the French Riviera, and eucalyptus trees in Madagascar.

When adding a new plant to the garden, it's important to know the potential risks.

THE RUBBER TREE

....

Hevea brasiliensis (A. Juss.) Muell. Arg.
EUPHORBIACEAE
Seringueira, Rubber tree

....

THIS IS A VERY BEAUTIFUL TREE. One of the biggest in the Amazon, it can reach sixty meters in height. Its trunk is splendid, but always scarred with grooves, for the *Euphorbiaceae* contains latex. One only has to wound a *Hevea* to see the white milk from which rubber is made ooze out.

If this tree intrigues me, it is because of its unusual rhythm of development. *Hevea* grows, then stops and loses its leaves before starting to grow again. The pattern isn't annual, but occurs over a period of forty-two days. Why does the rubber tree grow this way, as if it were governed by seasons—like European trees, but below the equator? It remains a mystery. Interrupt this rhythm, and the tree won't make latex. And if the latex is emptied out, the tree dies. No one knows what purpose the latex serves for the plant itself.

People primarily take interest in plants for what they can give them. For instance, we use medicinal plants without asking why it is that they synthesize such precious molecules in the first place. Everything has a function in the animal world; there, nothing lacks a purpose. The sheer number, among plants, of examples such as the rubber tree— whose latex has no clear reason for existing—makes me think they don't work like animals. Are they capable of selfless acts? The hypothesis isn't very satisfying, but that might be another thing that sets them apart from animals.

Europeans have known about latex ever since Christopher Columbus discovered America: he mentions it in his account of his travels. Rubber was brought to Europe for the first time in the eighteenth century by Charles Marie de La Condamine and François Fresneau de La Gataudière. The French naturalists were amazed when they saw Native Americans playing with a ball that bounced when dropped. La Condamine authored the first scientific description of the material, whose name in Quechua is *caotchu* (*cao*, "wood," and *tchu*, "which weeps").

In Brazil, the name for the rubber tree is *seringua*, and the man who harvests latex a *seringueiro*. Every night, the *seringueiro* leaves his hut and follows the same path: a circle through the forest. He bleeds each *hevea* along the way and leaves behind a coconut shell or a can to collect the latex. The tree is no hemophiliac, so it scars in a couple of hours.

At around eight in the morning, the *seringueiro* comes back home but almost immediately sets out again on the same route, this time carrying a bag into which he empties the latex that has collected in the interim. Once the second round is done, he needs to hurry and make a fire with palm fruit, the smoke of which helps shape the latex into large globes.

I've had occasion to admire the ingenuity of *seringueiros*—and Native Americans, for that matter—in finding practical uses for rubber. For instance: somewhere in the Amazon, a man riding a bike gets a flat. Undaunted, he removes the inner tube, looks for the right tree—needless to say, there's one around—and makes a cut in the bark. The latex oozes out, and he puts a little on the leak. The cyclist has just enough time for a cigarette before the patch is done; a few quick pumps, and the bike is back in operation.

Alternatively, a *seringueiro* might need a pair of boots. He fashions the shape of feet, right and left, with pieces of wood. Each day, he puts the molds into a latex bath, then lets them dry. A few days later, he's got a pair of boots!

But why do *seringueiros* always follow the same path in the forest? Why not grow a plantation of trees? Herein lies the great drama of rubber in Brazil: cultivating *hevea* is impossible because of a parasite—a fungus called *Hemileia vastatrix*. To avoid an epidemic, a distance of at least three hundred meters between trees is required. Under these conditions, intensive planting won't work. Henry Ford, the American industrialist, tried anyway in the 1920s. He had a whole workers' colony—christened "Fordlandia"—built on the banks of the Tapajós River. The thousands of trees that were planted soon perished because of the fungus, and today, Fordlandia is nothing more than a ghost town. Even a great captain of industry is no match for a mushroom!

The rubber tree claimed victory when it emerged from Brazil without its parasite. Brazilians maintain that the plant was stolen from them. They accuse the Englishman Henry A. Wickham of harvesting seventy thousand seeds with the help of Native American allies, tricking custom officials, and bringing the spoils to the director of the Royal Botanic Gardens, Kew, in London. There, it's said, the seeds were planted in tropical greenhouses, and the trees that grew were carried off to Ceylon (Sri Lanka).

But the story isn't at all true! Queen Victoria personally, and officially, asked for *hevea* seeds for Kew Gardens. Honored by the request, the Brazilians obliged. The first seeds didn't survive the long voyage by sail, as they only live for three weeks. It took a steamboat for new ones to reach England in a sufficiently healthy state to sprout in hothouses there. The rubber trees obtained in this manner then made their way to Malaysia, where plantations flourished—as they did in Ceylon and throughout Southeast Asia. Thailand is still the main producer of natural rubber.

Brazilians have forgotten the whole story and complain about botanical fraud. This development is understandable: it's painful for the native land of *hevea* to have to import natural rubber, and at considerable expense.

Still, the globalization of trade all but guarantees that the parasitic fungus *Hemileia vastatrix* will find its way to Southeast Asia. Cultivators are getting ready: recent plantings of *hevea* are no longer monocultures; instead, care is taken to follow the forest model and put at least three hundred meters between trees.

Today, even though most car tires are made out of synthetic rubber, natural rubber is preferred for high-performance products—tires for Formula One racers and airplanes, surgical gloves, and condoms ... On a philosophical level, *hevea* illustrates that what's natural will always win out over anything that's merely artificial.

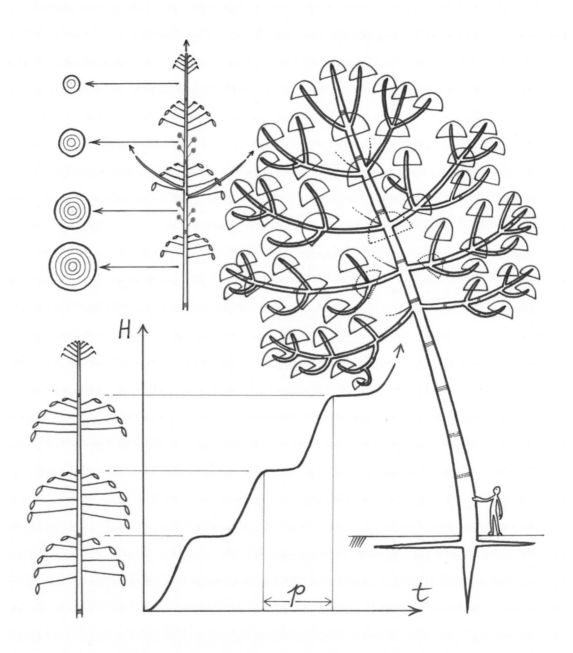

THE PLANT WITH JUST ONE LEAF

Amorphophallus titanum (Becc.) Becc.
ARACEAE
Titan arum

THIS PLANT is spectacular in more than one sense. Its inflorescence—which shoots up from an underground tuber—is as phenomenal as it is ephemeral. The native of Sumatra now draws tens of thousands of admirers to European botanical gardens where it's grown (in Nantes and Brest, for example). Visitors can also see it bloom at the Eden Project in Cornwall, southwest England.

It takes a few years for titan arum to flower, but when it does, it's awe inspiring. It makes a giant spathe two to three meters high that consists of countless microscopic flowers. It's one of the biggest inflorescences in the vegetable kingdom.

The flowers emit a fetid odor—hence the plant's nickname: "corpse flower." The putrid emanation starts when the temperature of the spadix increases dramatically. Meanwhile, a gigantic, purple spathe unfolds at its base. Beetles, especially ones that feed on carrion, can't resist all these charms: pollination of the *Amorphophallus* is guaranteed. The flowering doesn't last for long—just seventy-two hours. It yields pretty, red fruits that cover the plant, like those of the European arum.

> *The tuber sends up only one leaf a year, which is the size of a small tree: up to six meters high and five meters across. It's a leaf-monument, big enough for several people to find shade.*

Another curious feature is the fact that the tuber sends up only one leaf a year, which is the size of a small tree: up to six meters high and five meters across. It's a leaf-monument, big enough for several people to find shade. Each year, the old leaf dies and a new one takes its place. When the tuber has stored up enough energy, it enters a dormant period for about four months, then another bloom burgeons and the process starts all over again.

The titan arum is quite mysterious, to be sure. Its leaf is especially intriguing. Why is it always jagged? Maybe it's a defense mechanism: if it looks like it's already been gnawed at, potential predators will move on to find food that's still intact. To be honest, the hypothesis doesn't convince me, but it's the only one out there at the moment.

Common enough in Sumatra, this plant doesn't seem to be directly threatened. That said, deforestation is shrinking its habitat, and the International Union for Conservation of Nature has classified it as vulnerable. What's more, titan arum is often poached to be sold on the horticultural market in Korea and Japan.

THE BIGGEST TREE IN AFRICA

Baillonella toxisperma Pierre
SAPOTACEAE
Moabi

IN JUST ONE DAY in the Gabonese forest, one can encounter a good ten trees with a huge diameter, bigger than the room I'm writing in. But chancing upon a *moabi* is overwhelming. For me, this tree—with its stature, architecture, and longevity—is nothing short of a marvel. For centuries, it has dominated the ocean of the forest.

While filming *Il était une forêt* (*Once Upon a Forest*) in 2012 with the director Luc Jacquet, the crew had the good fortune of witnessing a *moabi* in bloom—a rare and fleeting event. Since there aren't any seasons in the humid tropics, trees that belong to the same species aren't synchronized. Accordingly, *moabis* don't flower at the same time, and the chances of catching one doing so are slight.

The *moabi* is the biggest tree in Africa, a giant that can reach seventy meters in height, with a four-meter diameter at the base. Dominating the canopy, it's what is known as an "emergent tree." It's difficult to determine the age of tropical trees—they often lack rings, and when they do have them, the rings don't follow an annual rhythm—but it is speculated that *moabis* can live for a thousand years.

Its imposing quality, its majesty, follows from its architecture—a rhythmical growth of ascending horizontal branches, with no reiteration whatsoever; it is beautiful, as are all unitary trees. Its thick, reddish-brown bark displays vertical furrows, like so many deep ravines. The bronze leaves of the *moabi* form clusters of stars at the end of the boughs. The flowers, which are small but abundant, have a pink tint that turns brick-red and emit a curious fragrance that calls to mind old, luxurious, and well-maintained homes.

The people of Gabon adore this tree. It plays a fundamental role in their life and culture, and it has given rise to many legends. For instance, the Pygmies claim that covering oneself with powder made from *moabi* bark makes you invisible. That said, I had no trouble seeing the Pygmy who told me that!

A large *moabi* can produce two and a half tons of fruit a year. When ripe, this fruit resembles a bulky pear, fleshy and golden. The seeds contain oil used for alimentary and cosmetic purposes. In this respect, I should point out that the scientific name for the *moabi* is poorly chosen. *Toxisperma* means that the seed is poisonous, but that's not true at all! Healers value the tree's bark for its medicinal properties.

The problem is that many animals like the *moabi*, too—especially elephants. The pachyderms pick up on the dull thud and low-frequency vibrations when fruit falls from some sixty meters up, and then they track it down by following its intense odor. Elephants contribute to the success of the *moabi*: when they eat the fruit, they swallow the seeds too, which they then disperse through their dung, miles away from the tree where they dined. It's also said that the seeds benefit from passing through the animals' digestive tract. When one of them sprouts, it yields, after ten years, a superb example of what's known as an "Aubréville model" of tree architecture.

But the role of the elephants isn't entirely positive: they also inflict deep wounds on the *moabis*. Looking for fruit when the time isn't right, they rush at the trees and stab at them. If the injuries extend around the whole circumference of a *moabi*, vertical transfers are interrupted: water can't make it up to the leaves, sap can't flow down to the roots, and the tree dies.

All the same, human beings pose the greatest threat to the *moabi*. The timber is perfect for woodworking and carpentry. During the colonial era, it was in great demand, and vast numbers of trees were chopped down and shipped to Europe. I spent twenty-five years working at the Botanical Institute of Montpellier, and all the doors and windows were made of *moabi*! Fortunately, a campaign conducted by Friends of the Earth, in association with village communities in Cameroon and Gabon, has borne fruit of its own. A number of large French businesses no longer import the wood, and Gabon banned the felling of *moabis* in 2010. Alas, Cameroon has not yet done the same.

For the time being, then, the tree is classified as vulnerable on the red list of threatened species established by the International Union for Conservation of Nature.

A PLANT
ON ROBINSON CRUSOE ISLAND

Gunnera peltata Phil.
GUNNERACEAE
Nalca, Pangue

THIS IS A SPECTACULAR PLANT, simply extraordinary, and seeing one makes a lasting impression. The first one I saw was on the Isles of Scilly, Cornwall, in the botanical garden of Tresco.

Subsequently, I had the opportunity to see others on the Juan Fernández Islands, an archipelago off the coast of Chile, on Robinson Crusoe Island. There, it is endemic, and vast expanses of them inhabit the slopes and floors of ravines. It has the appearance of a tree fern, with a soft trunk that can grow as tall as two meters. If pushed, the trunk collapses: it is not a tree, and it is not wood.

Up on top, gigantic leaves extend from a nest of ruby-red fibers; they're almost two meters wide. The foliage is really lush. These leaves are jagged, honeycombed, and lined with harmless spines. In the summer, lightly fragrant, pink flowers emerge at the top of the stem, on an inflorescence that can exceed a meter in length. The plant slowly expands as the stems grow and reach out; it takes ten years for it to reach maturity.

Gunnera looks like a giant rhubarb, but it isn't edible at all. Its stalk is loaded with bacteriochlorophyll—very green and capable of photosynthesis.

It took me three days to dissect one *Gunnera peltata*. Normally, a scalpel is used for dissecting plants. That time, I had to wield a meat cleaver!

It's an astonishing plant for the botanist, and I cannot think of a better way to present it than with a drawing. I will never forget my strolls through the *Gunnera* forest, beneath a roof of gigantic leaves.

THE BIGGEST FLOWER
IN THE WORLD

Rafflesia arnoldii R. Br.
RAFFLESIACEAE
Bunga patma in Malaysia

WE WERE WALKING in the Sumatran forest when, all of a sudden, a horrible smell overwhelmed us—like clogged toilets, or a garbage collector strike in the middle of August. This stench makes botanists laugh; for me, it calls to mind memories of the Asian forest. It means we're close to a *Rafflesia* flower, a parasitic plant discovered and described in 1818 by the English botanist Joseph Arnold, who named it in homage to Sir Thomas Stamford Raffles.

To find it, one must follow the path delineated by the pestilential odor. Then, there's no missing it. It's the biggest flower in the world: up to a meter across, thirty centimeters high, and weighing some seven kilograms. Needless to say, *Rafflesia* doesn't care about being discreet: it has five petals, thick and tough, and a crown of stamens and pistils the color of rotten meat. Given this smell and color, you can be sure that flies pollinate *Rafflesia*; there's always a cloud of them.

The flower is the only part of the plant that's visible. Its stalk and roots are inside the parasite's host: *Tetrastigma*, which belongs to the vine family (*Vitaceae*). The distance between the two plants can reach several meters; even so, the parasitic *Rafflesia* draws the water and nutrients it needs from the liana to which it's attached.

In 2013, *Rafflesia arnoldii* caused a minor sensation. At the San Diego Botanic Garden, one of them bloomed in a greenhouse. The staff made it do so by planting a *Tetrastigma*, then waiting until it was big enough to hold the weight of the flower. When the time was right, they made a cut in the vine and put *Rafflesia* seeds next to it.

Rafflesia poses a riddle. It rains a great deal in Sumatra, so how does the flower—which is shaped like a basin—manage to avoid overflowing with water?

THE BIGGEST LEAF
IN THE WORLD

Raphia regalis Becc.
ARECACEAE, OR PALM TREES
Royal raffia

I'VE ALWAYS BEEN interested in the biggest trees, their leaves, and their fruit … When I was a student, we were taught that the tree with the longest leaves in the world was *Raphia farinifera*. Its leaves measure twelve meters, which at the time seemed gigantic.

Many years later, in 1969, I was walking in the Forêt de Bangou with some Congolese and we came upon a palm. But all we really saw was the base of the leaves: their tops vanished in the canopy, high among the tall trees. My companions and I tried to bring a leaf down, but it took a good quarter hour of work with a machete to do so. Its fall was accompanied by an enormous noise—as if a whole tree were collapsing. We took the leaf back to the laboratory and measured and weighed it: twenty-eight meters long, four across, and a hundred kilos! The following month, we returned to Bangou and dug around the leaves to measure the buried part. The tree opened up and its leaves fell apart; these large *Raphias* only have seven of them, composed of long, lateral sections.

The tree was first described by Odoardo Beccari (1843–1920), the famous Italian botanist who specialized in palms. Its leaves are the longest in all the vegetable kingdom—twice as long as those of *Raphia farinifera*. My discovery made it into *The Guinness Book of World Records*!

Raphia regalis has another distinct feature: its imposing cluster of fruit, which is quite pretty, weighs some twenty kilos. The fruit's skin looks like a snake's hide, and if you give it a shake, you can hear the sound of something rattling inside. But once the tree has brought forth its mountain of fruit, its life is over. *Raphia regalis* displays the architecture named after the English botanist Richard Holttum: a unitary, branchless trunk; once it blooms, the plant is done, because there's no way out without any branches. This tree grows for dozens of years, flowers, and then dies. No wonder it's so rare!

Raphia regalis

Palmier acaule
sous forêt, sur sol de
coteau argileux très
bien drainé

Dioïque. (à vérifier).

7 grandes feuilles fonctionnelles
1 petite feuille axilant une inflorescence
+ bases de vieilles feuilles tombées

feuille :
longueur au dessus du sol : 22 m.
largeur ≃ 3 mètres
Circonférence basale : 50cm
folioles à peu près dans un plan.

infrutescence ♀
au moins 2 m. de hauteur.
(dessin un peu trop gros)

vu une infrutescence
de 3 m. de hauteur.
12/2/69.

"pétiole" d'environ 5 mètres.

forêt de Bangou, 26 Sept. 1968

THE FASTEST GROWING TREE IN THE WORLD?

Schizolobium parahyba (Vell. Conc.) S. F. Blake
FABACEAE, OR THE LEGUME FAMILY
Guapuruvu

I REALLY LIKE this Amazonian tree. It can be seen from far away thanks to its yellow-gilt bloom, which is absolutely magnificent. When young, it resembles a tree fern because of its enormous leaves, big enough to cover a table. In the dry season, when it flowers, it has no leaves at all and is completely yellow, and the forest floor beneath is covered in gold. It's one of the most beautiful trees to be found in the tropics.

When I was living in Ivory Coast, I obtained some of its seeds and planted them in front of our house. After a year, the tree had already reached nine meters; after two years, it was eighteen meters high, and after three, twenty-one. From the window, only the trunk could be seen—bright green like an apple and scored with marks made by its gigantic leaves. When full-grown, it stood twenty-four meters tall. A tree like that leaves a lasting impression!

Its wood is highly prized, and its amazing rate of growth makes it even more remarkable. In Brazil, there are *guapuruvu* plantations for lumber. A surprising feature of this tree is that it's gluey—if you put your hand on the young trunk, it sticks! That's why the top part is always covered with dead insects and looks like flypaper. The tree isn't carnivorous, though, and the bugs don't get digested. In fact, the "glue" comes out of the buds and keeps the growing leaves from drying out.

I met up with *guapuruvu* again in the Amazon. Its architecture follows the Rauh model, and is the same as that of a pine or an oak. In other words, all its axes—the trunk and the branches, too—grow upward; the branches occupy successive levels, corresponding to rhythmical growth; the top of the tree is covered with a bright expanse of yellow inflorescence, which makes it easy to spot when navigating the rivers.

Is *Schizolobium* the fastest-growing tree in the world? It is quite possible that someday botanists will find another plant in the same family that grows even faster.

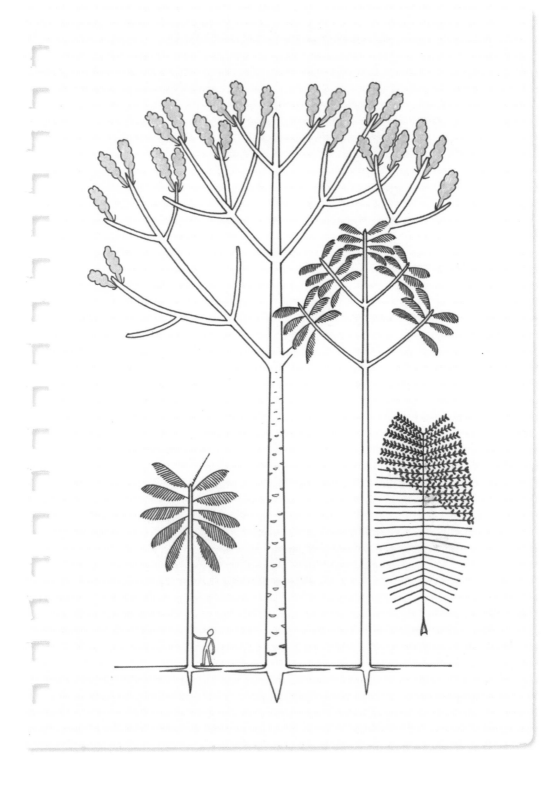

A GIGANTIC VINE

Entada gigas (L.) Fawcett & Rendle
FABACEAE, OR THE LEGUME FAMILY

—

IT'S SAID TO BE the longest plant in the world: up to a kilometer! But no one has ever managed to measure it precisely, because this Central African vine—I've also seen it in the Comoros Islands, in a second-growth forest—is hard to reach. It crawls up trees and grabs fast to the crowns of dozens of them with its hooks. It's so vigorous that it covers the heights of the forest, and so heavy it can bring down whole trees; when this happens, it just goes off to seek other victims in the vicinity.

However long it gets, *Entada gigas* can boast of having the biggest fruit in the world: its pods measure up to two meters in length and twelve centimeters in width. They hang down from sprawling vines, and to find them, you have to climb up the tree or pick pieces off the ground.

A single pod holds some fifteen flat, brown seeds that are shaped like a heart and quite pretty. They're often found on beaches, after rivers have carried them out to sea. Currents can take them from the coast of Africa all the way to South America; *Entada gigas* is also found in the Amazon forest.

What is the largest plant in the world, then? The tallest tree, the Californian sequoia, can reach 120 meters, but that doesn't make it the biggest by any means. A mere wisteria can span up to 152 meters—for instance, the one in Sierra Madre, California, which was planted in 1892 and now covers over 4,000 square meters (half a soccer field)! Each spring, this specimen brings forth more than a million pale, purple flowers. It's the biggest vine in the world to date.

Even though I'm one of the few botanists who bother with exceptional lengths, basal areas, or longevity, my reasons for doing so go beyond the kind of information that makes it into *The Guinness Book of World Records*. I don't care how many hot dogs somebody can eat, or how much beer they can drink, but I am interested in understanding the possibilities a given botanical species harbors: this is how our knowledge of the living world grows.

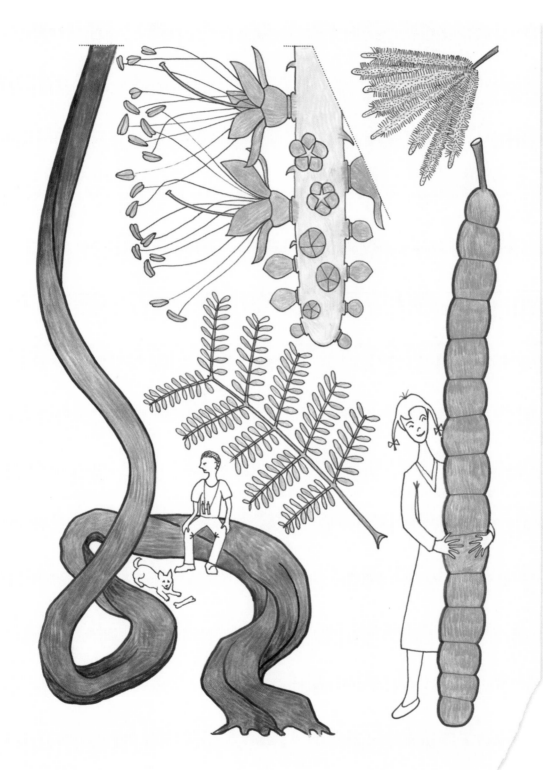

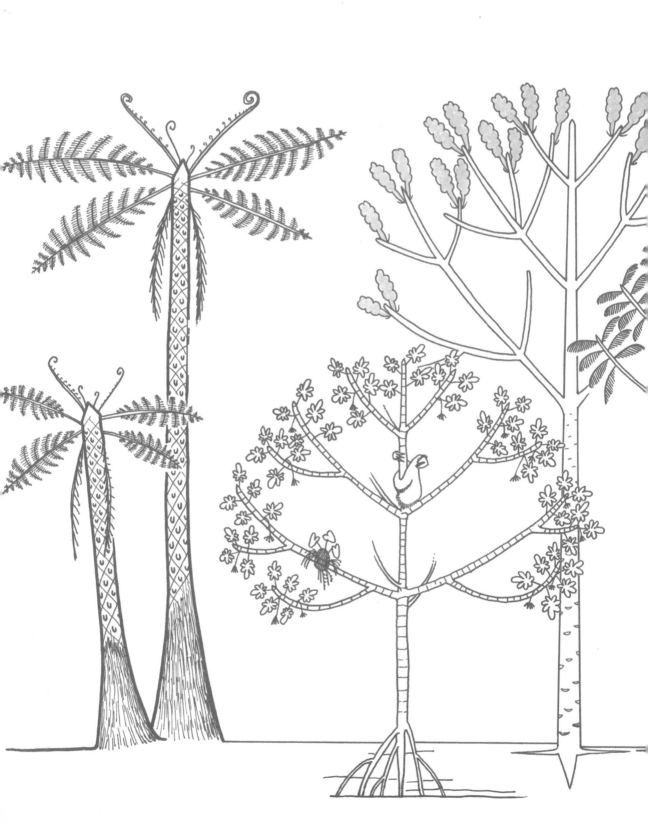

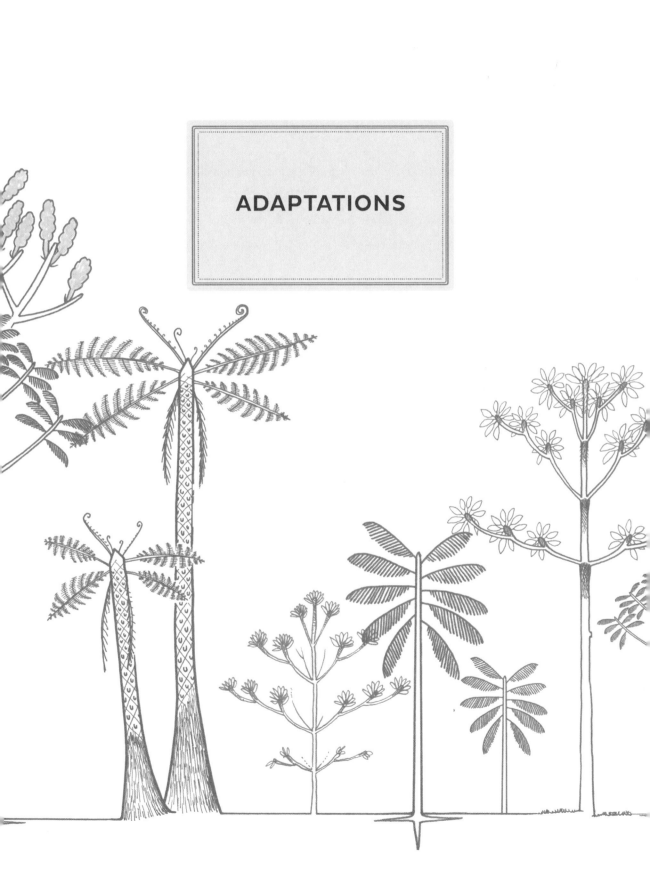

ADAPTATIONS

THE PLANT THAT IS
A SINGLE LEAF

....

Monophyllaea insignis B. L. Burtt—*Streptocarpus monophyllus* Welw.
GESNERIACEAE "MONOPHYLLAEA"
—

....

THE FIRST TIME I saw these *Gesneriaceae* in Africa, I was fascinated. Later, I rediscovered them in Borneo; there they proved to be just as intriguing. For a botanist, *Monophyllaea insignis* is one of the most surprising living things in the tropics. A single leaf growing on a substrate constitutes the entire plant, which can reach a meter in length: it really is strange.

It is found in dark, wet places—for instance, on the floor of old-growth tropical forests: *Streptocarpus* in Africa and *Monophyllaea* in Asia. There the plant belongs to those forms of tropical vegetation that have developed any number of stratagems for growing and reproducing in permanent shade.

Like all creeping plants, it is in constant danger of being covered up by all the debris—branches, dead leaves, and so on—that falls from the canopy. If this litter buries it, the lack of light can prove fatal. The only solution for these little plants, which are only a few centimeters tall, is to grow onto vertical surfaces where dead leaves can't build up. This is why they live attached to a cliff, a rock, or a tree trunk; they manage to do so thanks to countless tiny seeds.

That said, setting up shop on a vertical axis is difficult: there's no soil. So these plants keep their vegetative organization to a strict minimum, which is why they are monophyllous, composed of one big, hanging leaf with roots and flowers extending from the base (as the illustration shows). These *Gesneriaceae* are hard to miss.

Over time, the leaf expands at the base and deteriorates up on top. It's likely that these plants live for a few years, but their life span awaits further study.

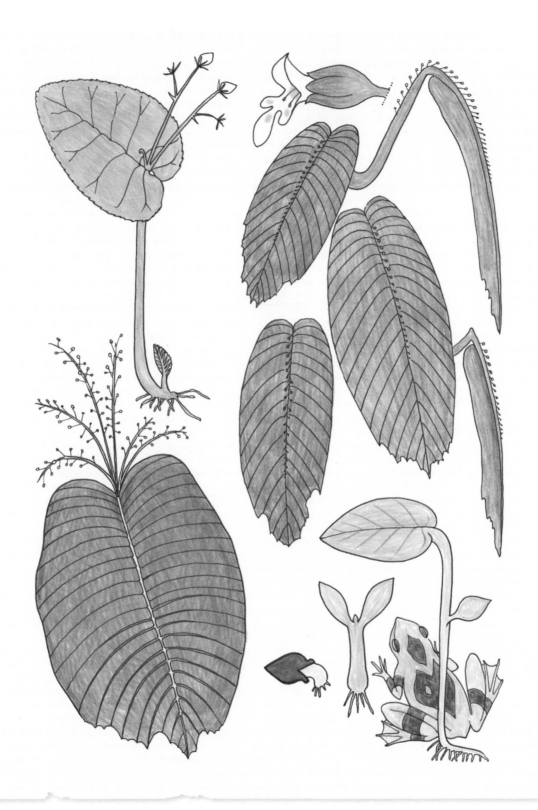

FLOWERS THAT GROW
FROM THE HUMUS

....

Duguetia calycina Benoist
ANNONACEAE
—

....

WALKING IN THE Amazonian forest, one sometimes sees a large, wine-colored flower on the ground—then two, three, four ... It's surprising, because in tropical forests it is more common to see flowers that have fallen from the canopy and broken up, whereas these shoot straight up from the humus. In old-growth, tropical forests, the floor is covered by a layer of carbon dioxide about fifty centimeters thick. If you tried to take a nap on the ground, you wouldn't be able to breathe. So how can a flower manage to bloom at ground level? To do so, it must be pollinated by something that isn't bothered by the carbon dioxide.

In the middle of this cluster of flowers is a small tree, with stilts at the base like the roots of mangroves. If you dig them up, you will discover that they're not roots but underground inflorescence—in other words, axes for making flowers and fruit. This is why *Duguetia* has flowers that shoot up out of the soil. The biology of these flowers and the reason they shoot up out of the earth are unknown. I should mention that they emit a fragrance that's pleasant and mildly alcoholic. They're probably pollinated by flying insects: diptera, hymenoptera, or butterflies.

I observed this phenomenon in the south of Cameroon, in a plant from a completely different botanical family, *Achariaceae*. That plant is named *Caloncoba flagelliflora* Gilg ex Pellegr. Walking in the rainforest of Campo, I came upon a carpet of tiny white flowers, poking straight up from the earth. It looked like a field of daisies; in the middle stood a little tree, about five meters high, with stilts at the base. When I dug them up, I discovered that they formed an inflorescence that could reach eleven meters in length.

I recommend that researchers interested in tropical rainforests conduct a study of the layer of carbon dioxide found there; its influence on life—plant and animal—at ground level remains unknown.

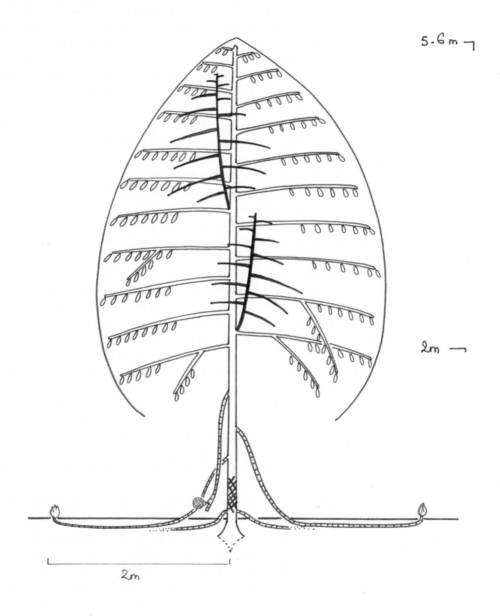

Duguetia friesii

F.H. n° 2632

ANNONACEAE

5.6m ⌐

2m →

2m

30 I 78

GUYANE. Sommet nord du Mt Galbao

THE PLANT DISGUISED AS A MUSHROOM

....

Helosis cayennensis (Sw.) Spreng.
BALANOPHORACEAE
—

....

IN 1962, when I was going up on La Mana River in French Guiana, I noticed what, at the time, I took to be a red mushroom with a cap that hadn't yet opened; the next morning, it was dead. Over the next few days, I saw many others, which all died just as quickly. Little by little, I realized that these mushrooms had all stopped living without giving me the chance to see them open up—as if something had happened to them during the night.

I decided to stay awake and watch, and I observed some very curious things. When darkness falls, the caps of these pseudo-mushrooms, which are composed of close-set scales, bring forth … flowers: little by little, the scales all fall away and a round cluster blooms. As the night progresses, these flowers wither and vanish, leaving a bare, red surface riddled with holes. Toward the end of the night, new flowers shoot out of the holes, and by four in the morning, the top is covered with little white blossoms that smell nice and attract insects. When day breaks, this second generation of flowers is already faded, and everything rots and falls apart.

What I'd taken for a mushroom was in fact a parasitic flowering plant. Living in the shady undergrowth, *Helosis cayennensis* has adapted to the absence of light and, therefore, of photosynthesis: it has no leaves or chlorophyll. Instead, it extends its roots into those of neighboring trees and draws energy from them, in the form of sugars and other products of photosynthesis. Over the course of the night, it first brings forth female flowers, and then male ones, which are the ones that smell

I myself don't care about finding a new plant that needs a new name. … My dream is to do away with the need for names and classifications and bring botany back to the study of the biology of plants.

....

nice. At dawn, the plant collapses. I was fascinated: I did not know that a flowering plant could look like a mushroom. I made a great deal of color drawings.

Helosis cayennensis inspires me to reflect on the practice of botany. Most of my colleagues dream of discovering a new plant. I myself don't care about finding a new plant that needs a new name. That's not my job. What interests me is the biology. My dream is to do away with the need for names and classifications and bring botany back to the study of the biology of plants.

To get back to *Helosis cayennensis*: it performs a legitimate type of parasitism, which, all in all, is perfectly moral. In tropical forests, the big trees enjoy the greatest part of the light up in the canopy. So down at ground level, the small plants—which are unable to perform photosynthesis in the shade—take their energy where they can find it: in the roots of the trees. It's only fair.

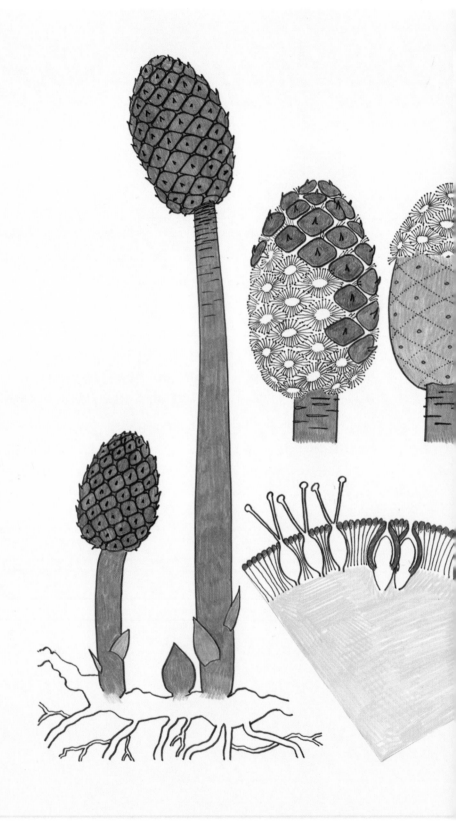

UNDERGROUND TREES

Parinari
CHRYSOBALANACEAE
South Africa

Jaborosa
SOLANACEAE
Argentina

WHAT IS THE DEFINITION of "tree"? When I say the word "tree," everyone can see what I'm talking about, but as soon as we try to define it, it proves to be nearly impossible. Does it mean wood? Branches? Is it a matter of having a single trunk, or can there be several? What about leaves? How do we provide a definition that leaves room for tropical and fossil trees? Even now, in the twenty-first century, there is no consensus over any definition of "tree."

Students at European forestry schools are given a definition that only covers trees in Europe, as if "our" trees—with a single trunk, wood, branches, a dense foliar crown, and an overall height of more than seven meters—were the only real ones. The oak, beech, and linden have nothing to worry about.

But such a definition covers neither fossil nor tropical trees. In the tropics, there are trees with a dozen trunks, not just one. Others, such as palms, have no branches or wood. Still others have no leaves—for instance, cacti or the large euphorbia found in Africa.

One day, I managed to formulate a definition that suited me—until underground trees were discovered in the savanna of South Africa, near Pretoria. My definition fell apart. It didn't work for such trees.

Students at European forestry schools are given a definition that only covers trees in Europe, as if "our" trees—with a single trunk, wood, branches, a dense foliar crown, and an overall height of more than seven meters—were the only real ones.

In the rainy season, all you see are big puddles of leaves and flowers, a few dozen square meters in size. Then, in the dry season, it all vanishes! One has to dig to find the vast, soft branches. Then there's no denying it: they are underground trees.

What does it mean to include a subterranean tree in a family that

48

isn't a family to begin with? Why would a tree bury itself? It's been remarked that these trees are quite ancient—some of them are over a thousand years old. It is also known that they grow where storms have caused fires in the savanna. It seems these plants adapted to deal with the results of lightning: blazes that have occurred regularly since time immemorial, long before there were any humans. Indeed, the ground provides effective thermal isolation that protects them. These underground trees are fascinating; fifteen species have been identified in Africa and America.

My English colleague Frank White expanded this notion to the forest as a whole in an article that caused quite a stir among botanists: "The Underground Forests of Africa: A Preliminary Review."[1]

1. *Gardens' Bulletin*, Singapore 24 (1976): 57–71.

?............

AN ORCHID WITHOUT LEAVES

....

Microcoelia caespitosa (Rolfe) Summerh.
ORCHIDACEAE
—

....

WHEN I LIVED in Abidjan, Ivory Coast, I worked in a garden where trees were covered in *Microcoelia*, a widespread kind of orchid in that country.

These plants are not parasites but epiphytes—which means that they grow on branches, far from the soil. For them to live at the tops of trees, three conditions must be met: their seeds must be as fine as dust, so the wind can blow them; there need to be lichens and mosses to retain some moisture on the branches; and finally, it requires a certain kind of mushroom, which orchids need to live. *Microcoelia* has no leaves or stalks. Instead, its roots are arranged in the shape of a star, in the center of which emerge pretty white flowers equipped with a long, transparent spur partially filled with nectar.

Plants that live without soil aren't made to hold water, so they need to avoid losing moisture. What dehydrates plants is evaporation that occurs through the leaves, so it makes sense that epiphytes often have small leaves. The foolproof solution is not to have any at all—like *Microcoelia*.

At the same time, the plant needs to perform photosynthesis, a function usually handled by leaves. How can it occur without them? *Microcoelia* has its chlorophyll in its roots. When it's dry, they're gray: a layer of dead cells surrounds each root. When it rains, this layer fills up with water, and the chlorophyll appears: the roots turn green. Moreover, the dead cells play another important role: they host the orchid's symbiotic fungus.

Microcoelia displays an altogether normal biology, then; it has simply replaced its leaves with roots. This adaptation is so practical, it never ceases to amaze me: "I don't have any leaves. Fine. I'll put the chlorophyll in my roots!"

Many orchids have no leaves, and they can be found in all tropical forests. One of them, in New Guinea, goes even further than *Microcoelia*: it has flat roots that take the form of leaves attached to the branch of a supporting tree!

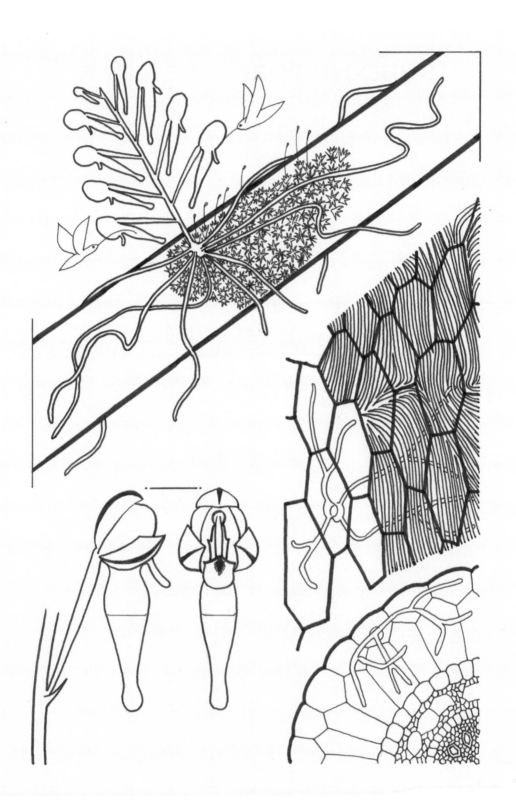

SUBEQUATORIAL GENTIAN

Voyria coerulea Aublet
GENTIANACEAE

—

IN THE GUYANESE forest, a carpet of dead leaves covers the ground. When walking there, one will notice, once or twice in the course of a day, a blue gentian—just like the ones in meadows in the Alps. Such an encounter, in the undergrowth of an equatorial forest, where only a little light filters in, is surprisingly moving. What's more, this gentian has a very pleasant smell.

Since it grows in the shade, making photosynthesis impossible, the plant has no chlorophyll; nor does it have any leaves. As such, it has to find its energy another way. By means of its roots, *Voyria* lives in symbiosis with a mushroom in the soil, which in turn lives symbiotically with the roots of a tree. The fungus absorbs the tree's sugary sap and transfers part of it to the gentian. In this way, the flower uses the energy the tree has captured.

One could call it poetic justice. *Voyria* is plunged into the shadows by the tall trees towering above that form the forest's canopy. To survive in the absence of light, it has to adapt by finding what it needs in the roots of these same trees.

Gentians without chlorophyll are also found in the equatorial forests of Africa, where they live in the same manner as *Voyria*. Needless to say, the alpine gentian lives in full sunlight and doesn't have this problem; it has green leaves, like other plants.

Such diverse modes of living—adaptations to constraints such as shade, lack of water, or the absence of soil—is what makes equatorial forests so fascinating. There is no question that researchers will find plenty of work on this terrain for generations to come.

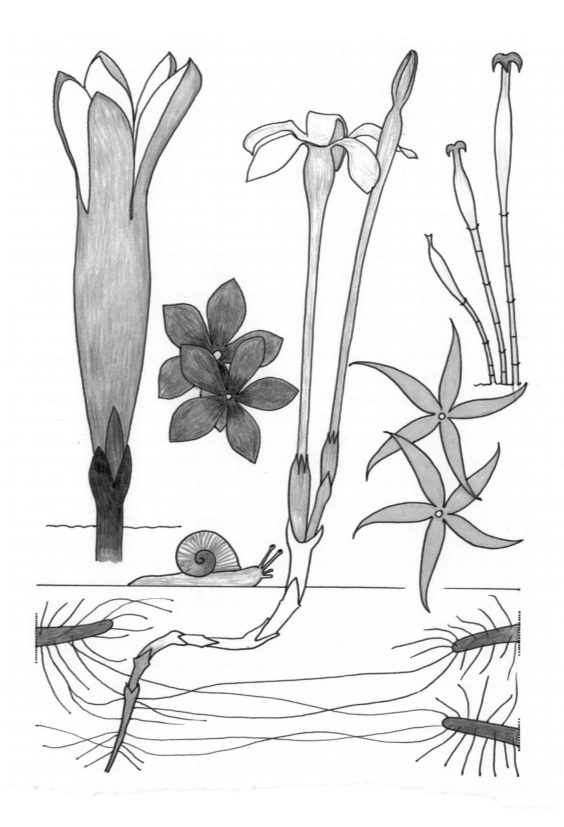

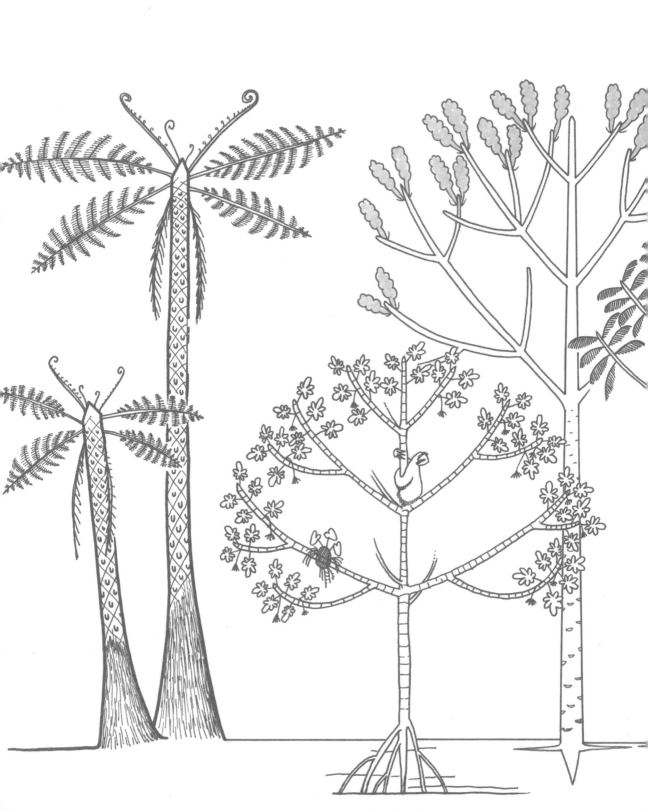

MYSTERIOUS
BEHAVIOR

THE TREE THAT CANNOT LOSE ITS LEAVES

....

Eucalyptus perriniana F. Muell. ex Rodway
MYRTACEAE
Spinning Gum

....

EUCALYPTUS ARE Australian trees; more than seven hundred species exist. *Eucalyptus perriniana* is just a modest bush growing in mountainous regions where it tends to be very cold. It is also known as the "spinning gum."

The juvenile leaves are round; they grow in opposing pairs, joined to one another. When they die, these leaves are unable to fall away. Instead, they slide down the stalk and make a pile at the base that acts as a kind of protective sleeve. Eventually, they decompose, but they do so much more slowly than if they had wound up on the ground.

People who see *Eucalyptus perriniana* at a botanical garden in Europe often laugh at this little tree that can't quite get rid of its dead leaves. But in the conditions where it normally grows, these sleeves may play a protective role against cold and snow.

At first glance—and for anyone unfamiliar with the species—*Eucalyptus perriniana*, the tree unable to lose its leaves, looks merely quaint. In fact, however, it has found an excellent means of surviving in the snowy and cold mountains that provide its native habitat.

A PARASITIC LAUREL

Cassytha filiformis L.
LAURACEAE
Love-vine, Devil's Gut

THIS HERBACEOUS LIANA belongs to the well-known laurel family, yet it poses a number of questions for biologists. In the south of France, there's bay laurel, whose leaves are used in cooking. But the family of *Lauraceae* exists all over the world, especially in the tropics. They are little trees with little green flowers and pretty leaves that are white underneath. Although their biology is unremarkable, *Cassytha filiformis* stands out in this understated family: it is unlike all other *Lauraceae*.

First named by Linnaeus in 1730, this skinny, yellowish vine crawls over its neighbors, especially trees. The parasitic plant holds fast with spikes that puncture the host's bark and reach vessels from which it extracts sap, nutrients, and water. The kind of host doesn't matter: it could be anything from a bush to a large fruit tree. The vine can cover a dozen victims at once and lose all contact with the earth. I remember seeing huge mango trees in Thailand that were covered by a *Cassytha*; they looked like they were wearing giant yellow wigs. Its filaments can devour them—a veritable plague!

But why does *Cassytha* live parasitically instead of practicing normal photosynthesis? Why isn't it like other *Lauraceae*? All families of plants include a maverick—a black sheep that doesn't do what everyone else does. Despite its strange appearance, there is no doubt that *Cassytha filiformis* belongs to the *Lauraceae* family: it has their characteristic anther and pollen. When a *Cassytha* seed sprouts, it initially yields a little plant that looks typical of *Lauraceae*. But as soon as it discovers a nearby plant to crawl on, everything changes: without leaving its family of origin, it stops being "normal," loses almost all its chlorophyll, and turns into a true parasite.

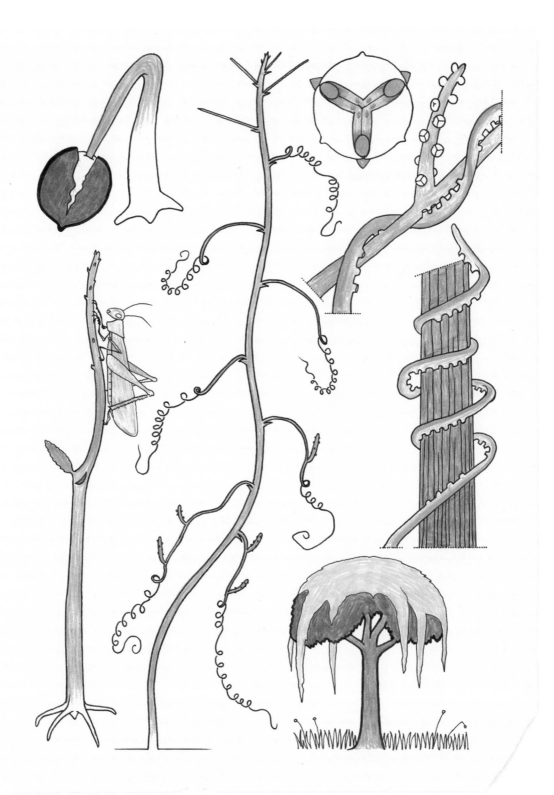

THE CHAMELEON VINE

....

Boquila trifoliolata (DC.) Decne
LARDIZABALACEAE
—

....

THIS IS A VOLUBLE LIANA that creeps, like wisteria or morning glory, by wrapping itself around another plant—not like ivy or sweet pea, with tendrils. Hoisting itself along the length of trees, this liana has developed a singular form of adaptation: to defend itself against herbivorous insects, it imitates the leaves of the supporting tree. Its leaf preserves three leaflets of its own, but otherwise it performs a complete metamorphosis to copy its host: it modifies its size, shape, color, orientation, and even the pattern of its veins in such a way that it fuses perfectly with the foliage of the tree that bears it. If, in the course of growing, it changes its support, the same stalk can even display leaves that are completely different, corresponding to the new tree—even if these leaves are much bigger. The surface of its leaves can change and get larger or smaller; it is able to modify the length of its petiole, too; the color can shift from bottle green to pale green, and it can become dull or shiny; it can even bring forth a little peak at the top of each leaflet. It's a real chameleon.

When it creeps up a tree with toothed leaves, *Boquila* does its best, but it's not very good at imitating this kind of plant!

Such imitation provides an excellent defense strategy against herbivores: weevils, snails, and leaf beetles. Chilean researchers have compared *Boquila*'s modified leaves attached to branches with those that are still young and crawling along the forest floor looking for a tree to climb. When the liana is still on the ground, its leaves get eaten; in general, predators spare the disguised leaves, which look like those of the host tree.

I've never seen *Boquila trifoliolata*. I only know about it from an article in a scientific journal that was published in 2014.[2] All the same, this liana intrigues me, and I'm not alone. Many biologists are eager for the chance to observe *Boquila*.

2.   E. Gianoli and F. Carrasco-Urra, "Leaf Mimicry in a Climbing Plant Protects against Herbivory," *Current Biology*, May 5, 2014.

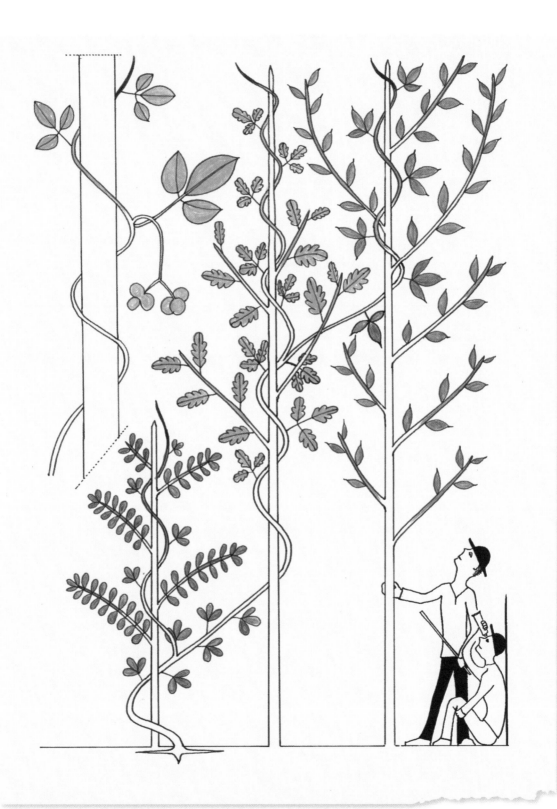

What's so remarkable about it? For one thing, mimicry, while common in animals, is rarely displayed by plants. One of the foremost examples of vegetal mimicry occurs in Europe: flowers of orchids belonging to the *Ophrys* genus have an imposing petal called the *labellum*, which looks so much like a female bee that male bees come and try to mate with it, thereby pollinating the flower.

Then we have another question: how does this chameleon vine manage to recognize, and then copy, its host's foliage? There are two hypotheses, but neither one convinces me. Researchers are inclined to posit a chemical or biological signal that the host emits naturally, which *Boquila trifoliolata* somehow manages to detect. This signal might be a volatile organic compound that triggers specific genes, which then modify young leaves as they grow. The second hypothesis relies on a "horizontal transfer," a bland term that in fact refers to a revolutionary concept in genetics. Customarily, sexuality is the means by which two living beings exchange genetic material. For animals, it's the only way. For plants, there's another possibility: "horizontal transfer" means that two plants that aren't related—say, a cypress and a leek—can exchange genetic material asexually. This

For plants, there's another possibility: "horizontal transfer" means that two plants that aren't related—say, a cypress and a leek—can exchange genetic material asexually.

occurs, for instance, when a plant-eating animal eats one plant, then another, thereby combining their DNA. Plants hybridized in this manner wind up resembling each other: either one mimics the other, or an intermediary form is adopted. But in the case of *Boquila trifoliolata*, it is not easy to see what animal would play the mediating role.

This liana is full of mysteries. We still don't know its biology well, and we're hard pressed to explain how it developed this singular form of adaptation.

What amuses me is that, according to official classification, *Boquila trifoliolata* belongs to a family—*Lardizabalaceae*—that includes the most primitive flowering plants, basal angiosperms. Although considered archaic in comparison to recent plants, dicotyledons and monocotyledons, it has developed a highly sophisticated mechanism of self-defense.

Boquila represents a new and particularly exciting case of information exchange between plants; my dream is to cultivate it in one of Europe's botanical gardens.

THE DANCING PLANT

Codariocalyx motorius (Houtt.) Ohashi
FABACEAE PAPILIONOIDEAE
—

MY FIRST TRIP TO CHINA was in 1989. CIRAD,[3] in Montpellier, asked me evaluate the botanical garden of Xishuangbanna in the southernmost part of the country, right next to the Laos border. The garden had been founded during the Mao era, when Europeans were not welcome.

I arrived on a Sunday morning, and right away, the director, Chen Jin, asked me: "Do you know the dancing plant?" He took me to some plants growing in pots, little shrubs that didn't really seem too impressive. Their leaves, which look like those of green beans, are gray. Although they have pretty, pink flowers, these plants appear unremarkable on the whole—modest and retiring. But Chinese people were sitting all around the pots, clapping their hands and shouting. "They're making the plant dance," Chen Jin explained. Indeed, the two lateral folioles of each leaf moved when sound was made.

"Could you sing a French song to see if that works too?" they asked me. A Breton sea shanty and ... it worked! To the immense delight of onlookers, the plant started dancing.

Before making the trip, I'd been told: "Watch out. The Chinese are intensely nationalistic—they don't want anybody pilfering their seeds." All the same, I couldn't help eyeing the fruits that covered the "dancing plants"—a profusion of little gray pods, like green beans. I eventually submitted a timid request: would the garden deign to give me a few seeds? Without further ado, the Chinese got up, went off looking for a big bag, and helped me collect some.

I took them back to France, where the plant can develop quite well in a greenhouse. The botanical garden at Les Cèdres, Saint-Jean-Cap-Ferrat, was the first place I grew one. The researchers here wondered if the movement occurred in response to

3.   Centre de coopération internationale en recherche agronomique pour le développement (French Agricultural Research Centre for International Development).

the air current produced by clapping or singing, so they put a transistor next to the plant. It danced without any problem!

Codariocalyx motorius makes it abundantly clear that plants intrigue us even more when they display animal characteristics and offer points of resemblance with ourselves. At the garden in Saint-Jean-Cap-Ferrat, the greenhouse featuring carnivorous plants is full.

What benefit does the plant derive from the astonishing ability to move its leaves when there is noise? It's still a mystery! If anything, I'd assume the opposite: that it would want to escape the notice of predators by not moving at all. Since my visit to China in 1989, our knowledge of *Codariocalyx motorius* has progressed, and researchers have brought to light a phenomenon I find quite amusing and surprising. When a specimen is left to grow for six months without any noise and efforts are then made to make it dance, it will move a little, but very slowly. But if it is trained every day, it will start dancing better and better, like a practicing ballerina. The plant needs training, in the athletic sense of the word, which must be based on some kind of memory.

When a specimen is left to grow for six months without any noise and efforts are then made to make it dance, it will move a little, but very slowly. But if it is trained every day, it will start dancing better and better, like a practicing ballerina. The plant needs training, in the athletic sense of the word, which must be based on some kind of memory.

We are far from solving the riddle of why this plant moves and how it does so. Needless to say, there is an explanation: it involves a complex mechanism that implies receptors for sound waves. No plant would just set up a mechanism like that for no reason, but we are still figuring out what that reason is. For all that, *Codariocalyx* has been known for quite a long time: Charles Darwin himself, in *The Power of Movement in Plants* (1880), sang its praises, and was quite aware that this plant moves. Science is often held to represent a cumulative process, but alas, this isn't the case: discoveries are made, which the next generation then forgets. Unfortunately, it happens all the time in botany.

At any rate, one thing stands firm about *Codariocalyx*: this plant senses sound, even though it has no ears.

THE WALKING TREE

Rhizophora mucronata Lam.
RHIZOPHORACEAE
Mangrove

MANGROVE FORESTS are the simplest and best-known tropical forests between land and sea, since few trees can handle the difficulties posed by saltwater and daily tides. The grayish silhouettes of mangroves, repeated ad infinitum, dot the landscape of low coasts in the tropics, where they perch on prop roots growing in the fine silt. The trunk and branches grow vertically, with no discernible rhythm; since its feet are planted in the seawater, the mangrove doesn't experience a dry season, so its growth doesn't follow any rhythm. To thrive in this soft, unstable soil, which lacks oxygen and is submerged at high tide, it has developed ingenious strategies: its leaves have the ability to get rid of the salt absorbed, and its aerial stilts give its roots room to breathe.

One finds two kinds of mangrove in the forest. There are those with a trunk that rises up from the soil; these specimens are sprung from a seed and don't move. Then there are mangroves whose trunk flattens out at the base; these trees stem from a branch at a low elevation—and they walk!

To understand what is going on, we have to look at how these trees start out. A mangrove seed sprouts and brings forth a little plant. As it grows, prop roots appear, which enable it to breathe and occupy a stable position on the mud; it's like a little tree on stilts. As the mangrove develops, more and more prop roots descend from its branches. At some point, they break off because they are no longer being fed by the trunk; now, the branches feed directly from their own roots. This is known as vegetative propagation by marcottage. The most striking thing is that, once detached from the trunk, the branches walk! Aerial photos taken at one-year intervals show that the branches move four or five meters a year. The bottom branch, which is always moving, heads for the sea, and when it can no longer reach the bottom—when it loses its footing!—it stops.

Is this tree really walking? When we walk, our physical mass moves as a whole. But no displacement of matter occurs with mangroves; the way they grow just makes it look that way. The branch experiences necrosis and vanishes at one end, while it keeps developing on the other. The "movement" is the growing process, four to five meters a year.

Mangroves form amphibious forests that are often quite dense, and create a natural barrier between the sea and the land. In tropical countries affected by cyclones, storms, and hurricanes, these forests are quite useful: they reduce the damage caused by winds and waves, even when a tsunami strikes. Mangroves also protect the coastal area from the erosion caused by swell and marine currents; they filter pollutants and heavy metals in the water; like all the world's plants, they trap carbon dioxide; and finally, they provide a habitat for important marine fauna.

Countries that have destroyed their mangroves for their tannins or for firewood regret doing so. Luckily, this kind of tropical forest is easy to plant and grows rapidly.

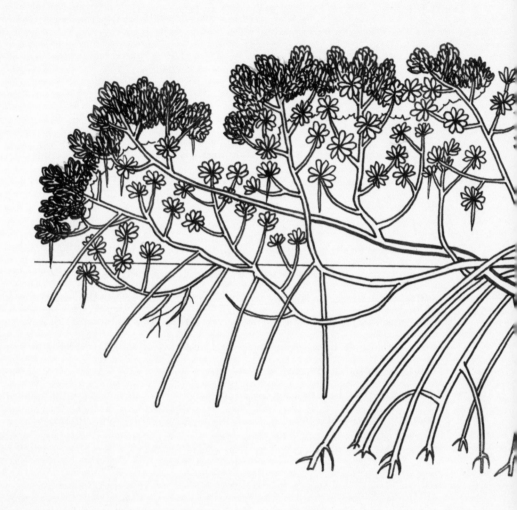

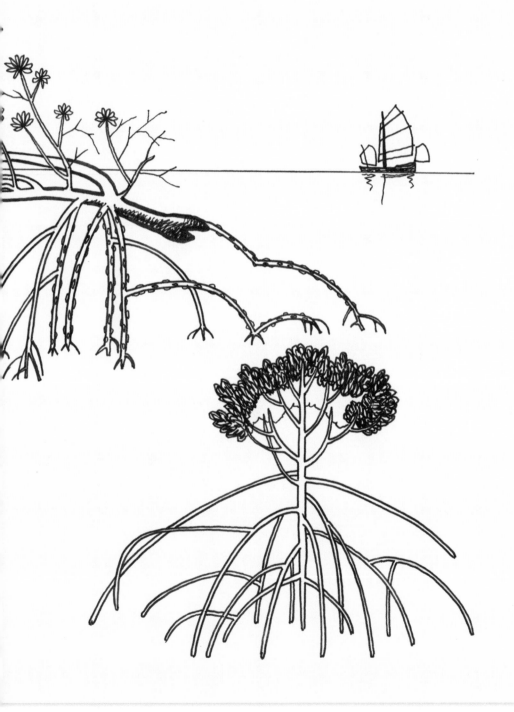

SPANISH MOSS

Tillandsia usneoides (L.) L.
BROMELIACEAE
Barba de Palo, Spanish Moss

TILLANDSIA is a bizarre plant: you can't tell its stalks from its leaves. They're just grayish threads of spindly organs—flexible, branching out in all directions, and covered with fine scales. It all hangs under a bough and looks like an old man's beard. This epiphytic plant, which belongs to the pineapple family, lives attached to branches exposed to full or partial sun, and it only thrives when the air is humid. Its appearance is truly astonishing. When it's not raining, it's gray; when it rains, it turns green.

Tillandsia usneoides has no roots. It draws water from what streams onto its stalks and leaves. The scales that cover it absorb water and minerals in the dusty moisture carried by the wind. Although it lacks roots, it has tiny flowers that are a lively, brilliant green.

Barba de Palo grows and propagates quite quickly, so it winds up blanketing whole trees. Birds use it to make nests; they carry off little pieces and put them on a neighboring tree, where they keep growing. This plant has a surprising quality: if there is no rain, it lives in slow motion. Like a moss, it lowers its metabolism, waits, and turns completely gray. When the rain returns, it turns green and starts up again. When organic processes slow down, respiration is quite weak; there aren't many plants that manage to pull that off. You can cut off a bit of *Tillandsia* and leave it in a drawer for a few months; then, when it gets some water and light, the plant will begin growing again. Needless to say, if you forget that it's there for six months, it will ultimately die.

This plant is quite popular in Latin America, where it is used for ornamental—and often amusing—purposes. Enormous clumps yield a dry but supple material that fills quilts, mattresses, and pillows; it is also used to make manger scenes at Christmas. At the beginning of the twentieth century, in Detroit, Michigan, *Tillandsia* received the honor of finding a modern, industrial application: seats in the famous Model T were filled with Spanish moss.

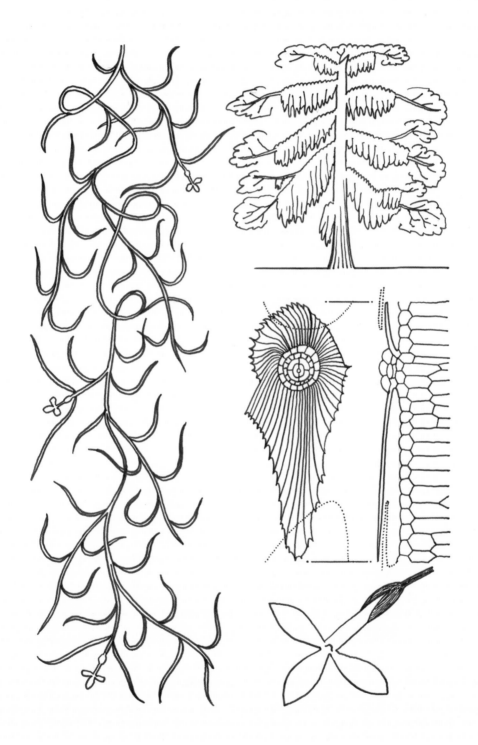

A PART-TIME CARNIVORE

....

Triphyophyllum peltatum (Hutch. & Dalz.)
DIONCOPHYLLACEAE
Airy Shaw

....

IT TAKES MILES of walking through the forest to find this rare plant, which is sometimes classified among the carnivores of tropical flora. When I was working at the botanical garden in Abidjan, I planted some seeds of *Triphyophyllum peltatum* under the mosquito netting I used as a conservatory. The plant started out by growing rather unremarkable leaves; but when it reached forty centimeters, its leaves looked like those of *Drosera*, the carnivorous plants of Europe. The limb is short and the midvein covered over with glandular bristles that secrete a red glue which insects mistake for nectar. In Europe, when prey makes contact with *Drosera*, it gets stuck and the plant digests it.

Thinking the same would hold for *Triphyophyllum*, I put flies on the leaves—but they didn't get eaten! Carnivorous plants grow in soil poor in nitrogen, so they get it where they can: from animal protein.

The hypothesis is that *Triphyophyllum* won't digest insects if it has enough nitrogen. When I planted it, the soil contained enough and the plant had no need of any insects. The plant is programmed to bring forth sticky leaves that alternate with normal ones, but it doesn't use them systematically: it's a part-time carnivore!

Triphyophyllum harbors other surprises. When it reaches fifty centimeters in height, it turns into a liana and starts creeping. To do so, it employs a third kind of leaf, which has two curved hooks that attach to support plants; in this way, it can make it to the top of the tall trees where it flourishes. Its flowers, white or pale pink, are discrete; the fruits resemble little flying saucers scattered by the wind. That's why it needs to spread out on the canopy, where there's wind. This liana grows by reiteration, in accordance with Oldeman's architectural model, with gluey leaves at the base of each segment. At our latitudes, the vine also grows by reiteration—like the wisteria entering my office through the window. Every now and then, I need to calm it down by pruning it—otherwise, it would colonize my desk, lamps, and books!

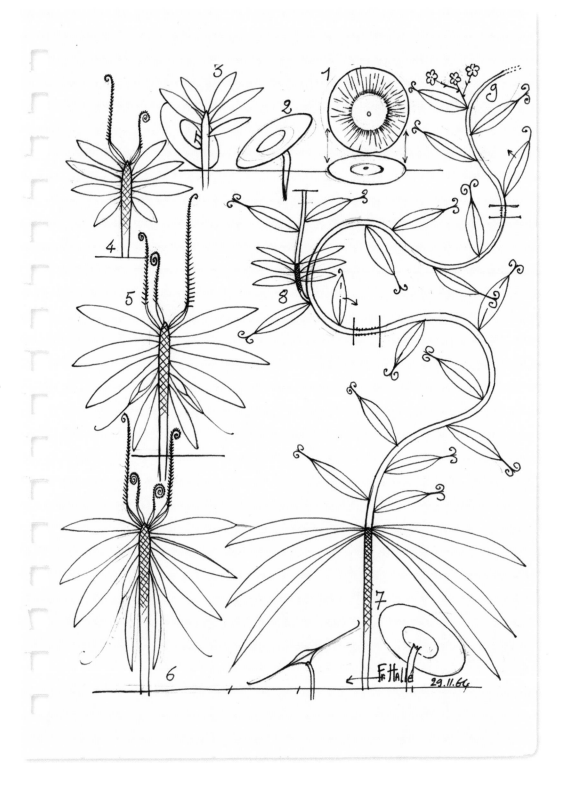

Fr Hallé 29.11.64

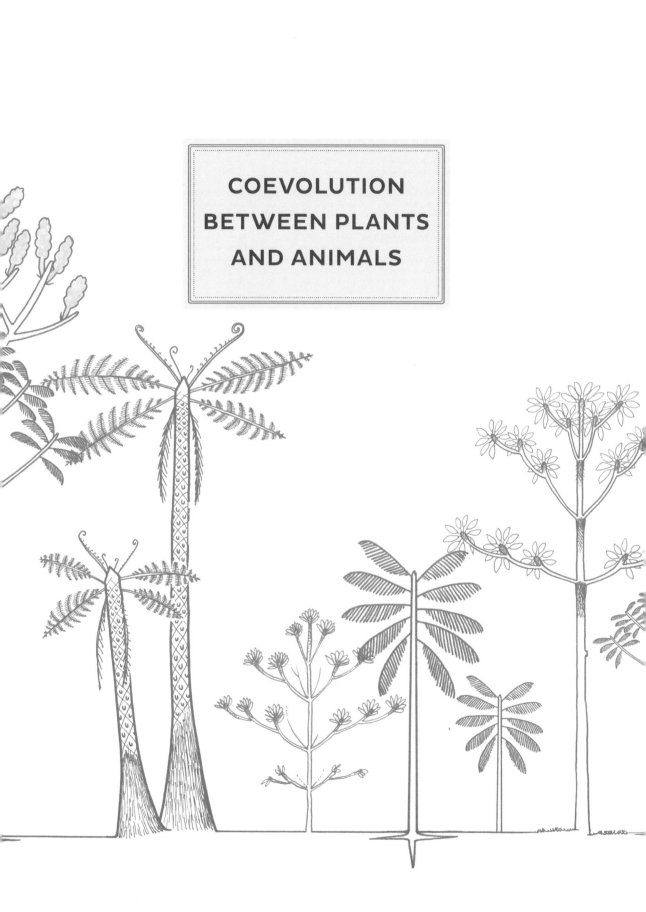

COEVOLUTION BETWEEN PLANTS AND ANIMALS

THE HUMMINGBIRD VINE

Marcgravia umbellata L.
MARCGRAVIACEAE
—

IN A SOUTH AMERICAN tropical forest, it's easy to find *Marcgravia*. These lianas are quite widespread. I saw them for the first time in Guadeloupe, on the slopes of La Soufrière, a volcano in the forest of Bains Jaunes.

When it's young, the vine crawls on the ground, looking for a support tree. When it reaches the base of the trunk, it uses its adhesive roots to climb up. At about three meters in height, it sends out branches, which expand horizontally, move away from the trunk, and develop brilliant leaves that are different from those of its immature stage. At the end of these stalks, hanging in space, appear striking flowers: round clumps with little vessels holding nectar on top.

The main stalk of *Marcgravia* keeps growing until it finally detaches itself from the support trunk. Then vegetative propagation starts by means of stolons—special runners without functional leaves, which emerge from the base, stretch out to the side until they collapse under their own weight, take root in the soil, and set out looking for new trunks to climb. In this way, *Marcgravia* manages to colonize a good ten trees in the vicinity.

Hummingbirds come and drink out of the little pitchers. At night, bats do the same. Ecologists have shown that *Marcgravia* has developed a strategy for capturing the bats' attention. The vessels for nectar send back a powerful, multidirectional echo—an acoustic signal—for the chiropterans that pollinate them, which sets the plants apart from the surrounding vegetation.

The fruit of *Marcgravia* is quite beautiful: under a brown husk that opens when ripe, there's a red placenta with black seeds; birds and monkeys love them.

With its crawling stalks and flowery axes hanging in the void, *Marcgravia* seems quite exotic. But in Europe, ivy grows exactly the same way.

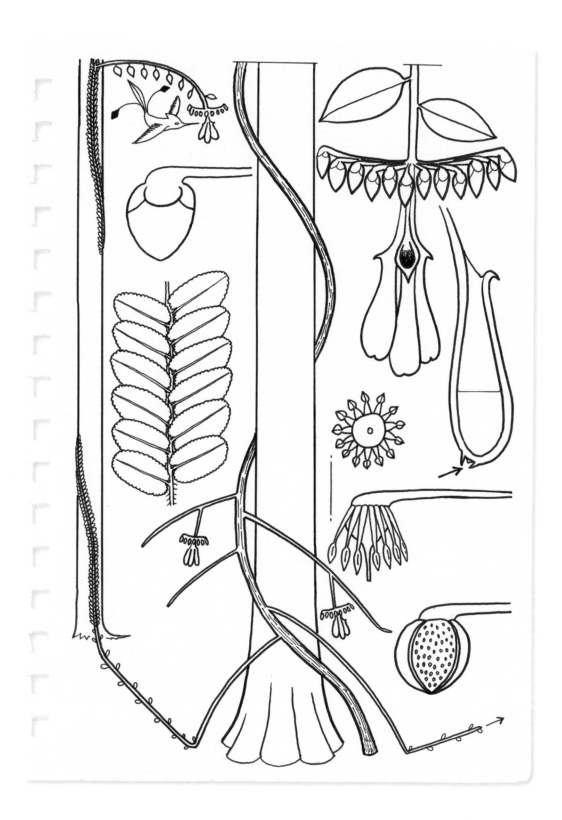

THE ADULTERY TREE

....

Barteria fistulosa Mast.
PASSIFLORACEAE
—

....

THIS LITTLE TREE is found at the side of the road, at the edge of the forest, and in clearings. In Gabon, it's common—and dreaded, too. *Barteria fistulosa* at first seems relatively unremarkable with its straight trunk and horizontal boughs following no particular order, even though, like all passionflowers, it displays a beautiful white bloom. But its branches, which are hollow and cylindrical, are home to countless *Tetraponera* ants: these large, flat insects are quite aggressive, and their stinging bites can lead to fainting or even death. The danger they pose keeps even the monkeys at a respectful distance. This little tree provokes so much fear that it's hard to find anyone willing to cut it down!

From the tree's own point of view, living in symbiosis with the ants offers an incomparable advantage, for they provide excellent security against predators. Any creature that comes into contact with *Barteria fistulosa* is immediately attacked with bites and stings. If a human being comes within a few dozen centimeters of the trunk, the ants drop from the branches and strike. In exchange for this protection, the ants enjoy the flowers' nectar and find shelter inside the hollow branches.

The local population has long known about the aversion these ants have to people. In the past, in Gabon and Cameroon, *Barteria fistulosa* was enlisted to punish women for adultery: the guilty party, made to stand at the trunk, immediately incurred the wrath of the ants. No one has ever told me of a comparable torture for men.

ONE OF THE PINNACLES
OF PLANT–ANIMAL COEVOLUTION

Cecropia peltata L.
URTICACEAE
Imbauba (Brazil), bois-canon (Guyana)

THIS TREE WITH PROP ROOTS belongs to the nettle family, but it doesn't sting. Did it lose its ability to do so, or has it not developed it yet? To make up for this lack of defense against predators, it lives in symbiosis with *Azteca* ants, which are quite combative. Getting stung by a nettle hurts much less than the pain *Azteca* ants inflict!

At first, *Cecropia peltata* grows without its partners; then, after a few months, it starts making fake ant eggs at the base of its leaves. Their shape, weight, and biochemistry are perfect, which proves impossible for the *Azteca* to resist. The ants rush in to collect the eggs; since the trunk and branches of *Cecropia* are hollow, it doesn't take much work for their mandibles to penetrate the tree's bark. Without further ado, a queen sets up shop and lays her eggs.

Such symbiosis represents a pinnacle of coevolution between plants and animals: the ant is sheltered, and the tree is defended. If a predator tries to get at the leaves, the aggressive ants swarm out by the thousands and chase it away. For instance, if a butterfly lays its eggs here, the ants will shove the caterpillars to the ground. And any botanist who tries to take a sample will immediately get stung! Sometimes, one finds young *Cecropia* with holes in their leaves: these specimens haven't managed to win over the ants, and they won't survive for long. Curiously, *Azteca* ants spare the sloth—the great tree-dwelling mammal of South American forests. Is this because the sloth is covered with thick fur that offers no weak spots? At any rate, the animal feeds on *Cecropia* leaves.

Sometimes, other ants make nests in these trees; surprisingly, the *Azteca* don't bother them. These other ants are small and inoffensive; they chew vegetable matter, which they use to make a kind of cardboard for building their homes. This material includes seeds that readily sprout, yielding an "ant garden."

How do these plants—which are always the same kind—benefit the little ants? It's not easy to get them to talk, but we know what's going on with some of the plants they "grow." For instance, the large leaves of *Araceae* play the part of umbrellas, which is quite helpful when your nest is made of cardboard and there's tropical rain. The roots of other plants form the walls of the nest; still others provide food. These nests don't last long, but the "ant gardens" keep growing until they form giant tufts on the upper branches of *Cecropia*.

Bats love the fruit of this tree, and they eat it in flight. It takes a few minutes for the fruit to pass through their system, and they also defecate in flight. The forest is thus constantly bombarded by bat dung full of seeds. A further point of distinction for *Cecropia peltata* is the vast quantity of seeds it produces, which can be found everywhere in the soil, about twenty per square meter.

These seeds sprout, but there are too many of them for tall trees to result. From a human perspective, the process of self-selection is pretty terrifying and amoral. As holds for all species, some individuals have better genes than others: they grow faster and take over the root systems of the slower ones. Up above, the specimens with less vigorous genes die, but their roots stay alive and capture water, to the benefit of the surviving plants. In this way, all the water absorbed by the weak passes to the strong. With each generation, the most fragile plantlets are eliminated; the robust ones hybridize and prepare the next generation. It's easy to see why *Cecropia* are thriving!

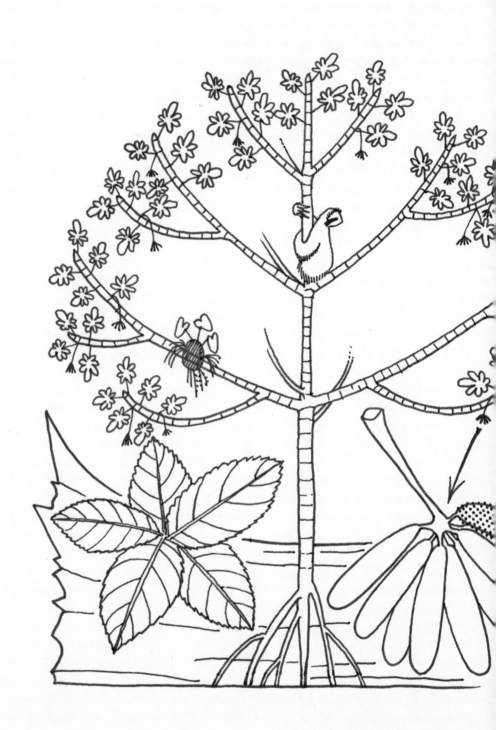

THE HANGING AQUARIUM

Guzmania lingulata (L.) Mez
BROMELIACEAE
—

"HANGING AQUARIUMS" are one of the marvels of American tropical rainforests. These epiphytic plants are found on all the great trees here, in full light, up at the top. From down below, it's hard to see *Guzmania*—and impossible to imagine what's happening inside. The plants grow on branches thanks to the wind, which carries countless seeds as tiny as specks of dust.

The leaves stand so tight together that they form an impenetrable reservoir for rainwater. Here, in a forest where no three days pass without precipitation, *Guzmania* resembles a rain gauge. With each storm, the plant fills up, holding as much as twenty liters of water. The water disappears by evaporation—which doesn't take long under the full, tropical sun. Thanks to the rain, this epiphytic plant keeps a major supply of water at the ready, even though it's far from the ground.

Exposed to sunlight, a whole ecosystem develops inside each hanging aquarium: frogs, mollusks, shrimp, insect larvae, and a kind of crab found only in these *Bromeliaceae*; it even hosts another carnivorous aquatic plant, *Utricularia reniformis*. At the base of *Guzmania*, the dead leaves make compost, where other animals live: scorpions, millipedes, cockroaches, termites, and even a blind snake of the *Geophis* genus.

Guzmania lingulata harbors another mystery. Some researchers think that when the plant blooms, it turns carnivorous and kills off its fauna for the benefit of its own inflorescence at the bottom of the aquarium. I'm not so sure: that would be one of the few examples of a carnivorous monocotyledon. To be absolutely certain, one would have to study this plant in a European greenhouse and analyze the biochemistry of the water when it blooms. Finding a digestive enzyme would prove that it had turned into a carnivore. Such analysis isn't easy, though: the plants are perched atop tall trees.

These hanging aquariums demonstrate that the canopy is the most extraordinary habitat in the tropics; exploration of the ecological and biological diversity here has hardly even begun.

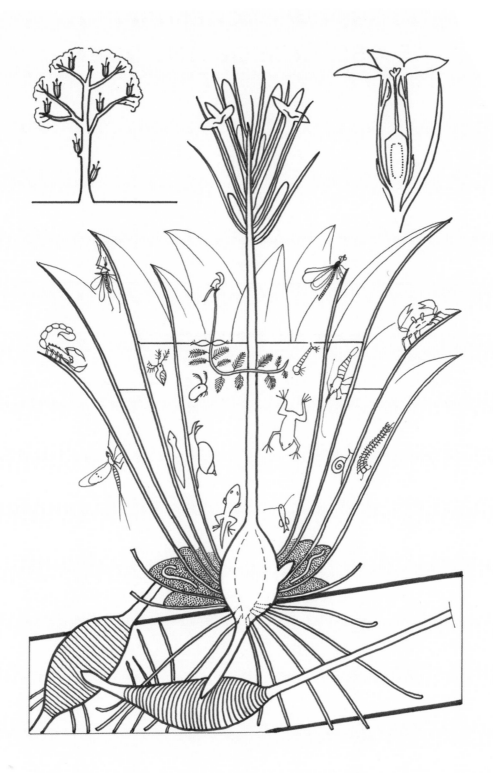

SHREW EXCREMENT AS A SOURCE OF NITROGEN

··

Nepenthes lowii Hook. f.
NEPENTHACEAE
—

··

NEPENTHES IS A CARNIVOROUS LIANA that displays originality both in its system of capturing prey and in terms of its relationship with insects and small rodents.

Nepenthes was observed for the first time in 1658 by Étienne de Flacourt, the French governor of Madagascar. By his account, it's a strange plant that bears, "at the end of its leaves, flowers or fruits that look like little vases."[4]

Linnaeus gave this tropical plant the name *Nepenthes* in 1737; it refers to a passage in *The Odyssey* where Helen pours the drug "nepenthe" into wine to make her guests less alert. When he chose the name, Linnaeus wasn't entirely convinced that the plant was carnivorous. He thought the vases contained only rainwater. Its carnivorous nature was later demonstrated by experiments performed by Charles Darwin and Joseph Hooker, the director of the Royal Botanic Garden, Kew; the results appeared in 1874 in the *Gardeners' Chronicle*.

Nepenthes grows in soil that's poor in nitrogen—for instance, the embankments of roads or at the tops of mountains. Since these plants need nitrogen to grow, they have developed "insect traps." Indeed, feeding on animals is the simplest way for them to get nitrogen.

At the end of each leaf of *Nepenthes*, there's a long tendril bearing a large vessel that's initially closed by a tight cover; later, when this lid opens, the vessel turns into a slippery and deadly trap, with a kind of digestive juice produced by glands at the bottom. It's full of dead insects: ants, spiders, flies, and wasps drawn to the magnificent red at the opening only to fall in and meet their doom.

4.   *Histoire de la grande isle Madagascar* (Paris, 1658/1661).

In the mountains of Malaysia, at an elevation of over two thousand meters, *Nepenthes lowii* Hook. f. has adopted a different strategy.[5] Its vases resemble a toilet bowl, with a sturdy stalk and a vertical cover that remains largely open; this makes the aperture much more capacious than those of other kinds of *Nepenthes*. Inside, in the digestive liquid, animal droppings can be seen. The plant's cover has little glands that secrete sugary fats that *Tupaia montana*, a shrew, absolutely loves. It's funny: the greedy shrew sits and puts its snout on the lid; while it's eating, it also relieves itself—thereby giving the plant the nitrogen it needs. The higher the altitude, the fewer insects there are; *Nepenthes* has found another way to get nitrogen!

How did this adaptation come about? How can a plant change its source of nitrogen? We still don't know. Examples of sophisticated adaptation like this reveal just how little we understand the plant world, even now.

5.   Charles M. Clarke et al., "Tree Shrew Lavatories: A Novel Nitrogen Sequestration Strategy in a Tropical Pitcher Plant," *Biology Letters* (2009).

At the end of each leaf of Nepenthes, *there's a long tendril bearing a large vessel that's initially closed by a tight cover; later, when this lid opens, the vessel turns into a slippery and deadly trap, with a kind of digestive juice produced by glands at the bottom. It's full of dead insects: ants, spiders, flies, and wasps drawn to themagnificent red at the opening only to fall in and meet their doom.*

PLANT OR ANTHILL?

Myrmecodia
RUBIACEAE
"The Devil's Balls"

IN 1999, on an expedition directed by Alain Conan, I participated in a dive to explore what remained of the *Astrolabe* and *Boussole*, ships that had sailed under Jean-François de Lapérouse; they sank in 1788 on the barrier reef of Vanikoro, in the Salomon Archipelago, north of New Caledonia.

I had the same role as the botanist who accompanied Lapérouse: Joseph La Martinière, a young man of twenty-seven who dreamed of following in the footsteps of naturalist explorers such as Antoine Laurent de Jussieu, Joseph Pitton de Tournefort, and Pierre Poivre. It was my job to collect plants on the island. La Martinière had amassed great quantities of them on his tour of the world, but they all went down with the ship. He'd also brought along the seeds of European plants to try to make them grow in the new environment. Have some survived on Vanikoro? I didn't see any.

The trunks salvaged on the expedition didn't contain an herbarium, needless to say, but we did find navigating instruments, thermometers, and silverware bearing the arms of the captain of the second ship, Fleuriot de Langle. I made a drawing of them, and we sent it to one of his descendants, an elderly lady living in Hyères; since she didn't have any heirlooms from her ancestor, she was quite moved.

Collecting specimens on Vanikoro, I encountered *Myrmecodia*: odd tubers hanging in abundance from the tree branches of the coastal forest. I asked my Melanesian companion what these plants were called. "Ah, sir, with all due respect, they're the Devil's balls!" he replied.

I already knew that about this variety of *Rubiaceae*, a classic in works of tropical botany. Examples are found in many European greenhouses—for instance, the Botanical Gardens in Nantes and Brussels. It's an epiphytic plant, that is, it lives attached to a tree; it's not a parasite, however, because it doesn't harm its host in any way.

Myrmecodia doesn't sprout on the ground, but it can grow a few meters up on any tree. Its tubers look like potatoes, and they also have their color, consistency, and taste. If you cut one open, you realize that it's a miniature colony. Harmless little ants run around in the passages hollowed out in the flesh; the ants are so small that they wouldn't be able to withstand the island's torrential downpour. The openings of their residence face downward, which keeps the rain from getting in.

Some galleries on the surface are designed for ventilation: there, the ants take care of their eggs and larvae. Others, which are deeper down and terminate in a dead end, have walls lined with "warts"; there, dead ants and excrement accumulate, yielding compost that's rich in minerals and nitrogen. The role the ants play is not to defend the plant, then, but to help it grow with the composting material: the warts are specialized in absorption. Drawing its energy and minerals from within is strange: *Myrmecodia* seems more like an animal than a plant!

Are Devil's balls plants or ant colonies? The question barely makes sense, as the one couldn't live without the other. This is true symbiosis. The ants provide fertilizer, and the plant gives them housing; they don't need to make the passages: the plant produces them on its own, and the ants just move in. If the plant is grown in a European greenhouse, the seeds sprout and make the hollow galleries without any ants; that said, the plant remains small and develops poorly, because there's no internal compost. It's a model case of coevolution between plant and animal. But *Myrmecodia* poses one of the most profound questions of the equatorial forest: what interactions, in the past, tied the ants to the plant, such that it now has the structure of an anthill—and even, in a certain sense, of an animal? Can genetic exchange account for vegetal life that adopts animal characteristics?

Clearly, the forest of Melanesia still has many strange things to teach us, including matters that bear on the most fundamental issues of biology.

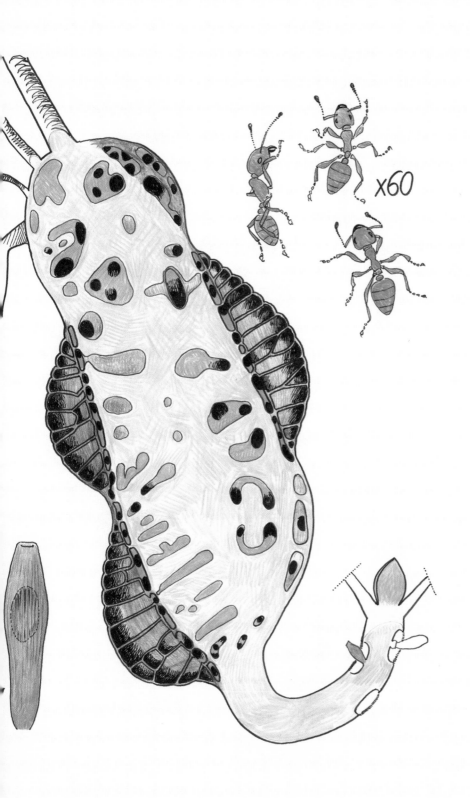

x60

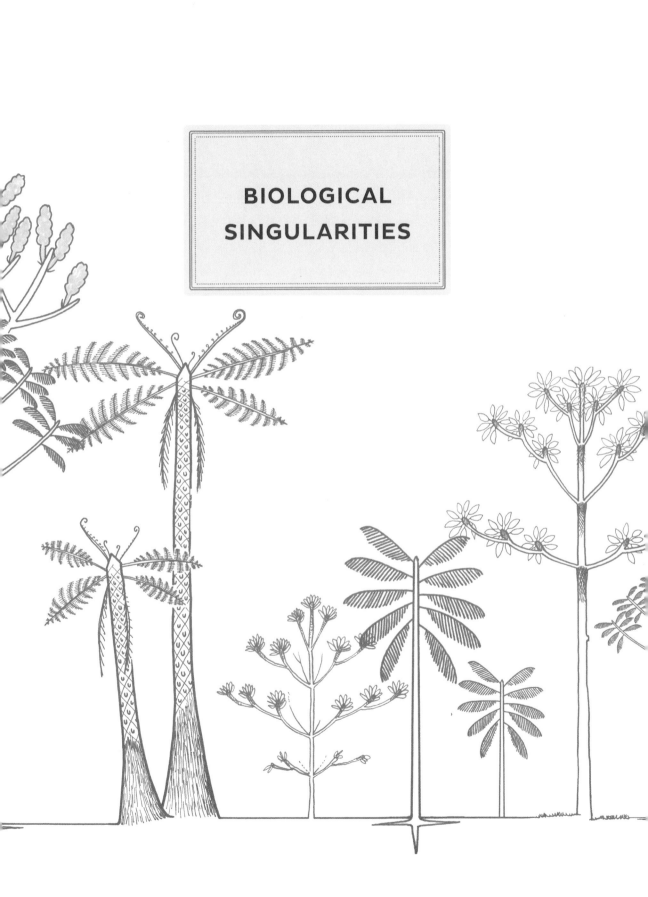

BIOLOGICAL
SINGULARITIES

A SURVIVOR FROM THE JURASSIC

Araucaria Juss.
ARAUCARIACEAE
—

WITH THEIR PERFECTLY regular architecture, gigantic stature—they can grow up to eighty meters—and the somber color of their crowns, which are almost black, these conifers of the southern hemisphere are astonishingly beautiful. Tall, straight, and tightly grouped, they make a real impression.

Araucaria is a unitary conifer, that is, it ascends vertically without reiteration. Such simplicity—its matter-of-fact geometry—is what makes it so majestic and commanding. "Modern" trees don't grow this way. *Araucaria* bears witness to an ancient period of the Earth's history: it's an archaic survivor from the age of dinosaurs, the Jurassic (150 million years ago).

Besides looking good, some *Araucaria* display a truly unique silhouette: for instance, in New Guinea, *Araucaria hunsteinii* K. Schum. can stack two, three, four, or even five crowns on top of each other. These crowns grow at the base of the tree and move up as the tree develops. The branches form levels; each year, another one is added, which makes it easy to calculate the tree's age.

When the voyagers saw these trees in the distance, they were confused by their unusual shape. "We could not agree in our opinions of what they were," the captain noted. They took them for giant basalt pillars before realizing they were pines.

Of the nineteen species of *Araucaria* on record throughout the world, thirteen are endemic to the archipelago of New Caledonia—including the Cook pine, *Araucaria columnaris* (G. Forst.) Hook., discovered in 1774 by Captain Cook and the Forsters, a father–son team of botanists. When the voyagers saw these trees in the distance, they were confused by their unusual shape. "We could not agree in our opinions of what they were," the captain noted. They took them for giant basalt pillars, then for pines. In due course, they christened the island the Isle of Pines.

Now, in western France and England, it is easy to find examples of one of the loveliest *Araucaria*: the Chilean species *Araucaria araucana* K. Koch. The Bretons call it "monkey puzzle" because of the spiny leaves that cover the trunk and make it impossible to climb. The name makes Chilean foresters laugh, because there aren't any monkeys in Chile! I'll never forget the sight of a forest of *Araucaria araucana* on a Chilean mountain during a tempest: the trees resembled a flock of huge, black animals covered with snow, huddling close to wait out the storm.

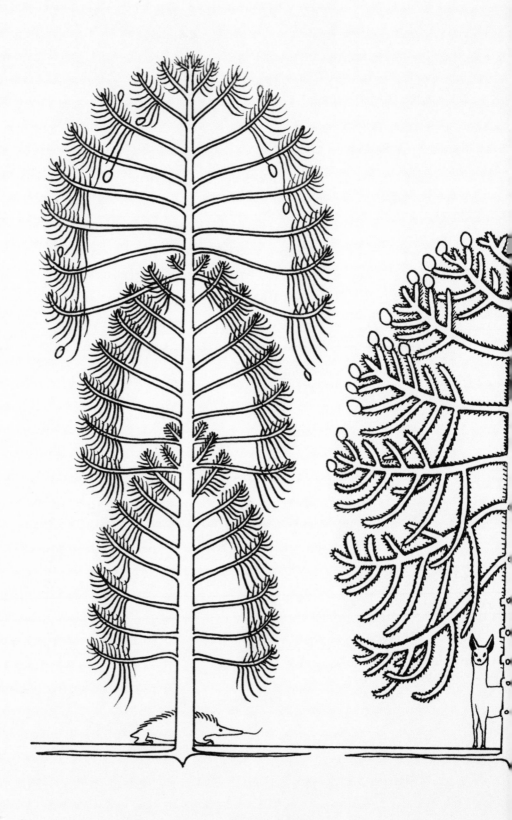

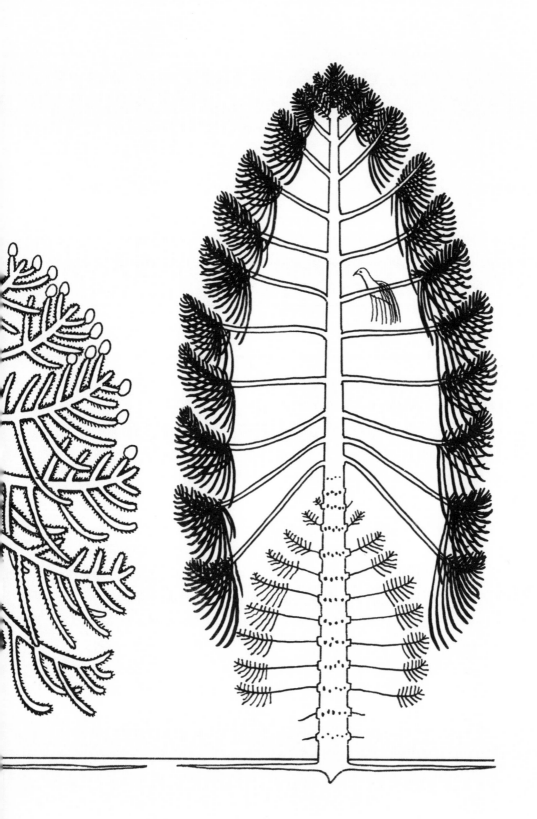

THE FOUNTAIN TREE

....

Ocotea foetens (Aiton) Baill.
LAURACEAE
Garoé, fountain tree

....

ON EL HIERRO, a small, dry island in the archipelago of the Canaries, it almost never rains, and for more than two hundred days a year, a sea of clouds covers the highlands. But there is a tree capable of capturing water from the fog: the *garoé*. It resembles a laurel, and when the moist air passes through its leaves, condensation occurs and drops of rain fall to the ground. It rains under the tree!

The first mention of this phenomenon dates from the sixteenth century, when the conquistadors laid claim to the Canary Islands. In his *History of the Indies* (1559), Bartolomé de las Casas wrote: "A little cloud always hangs at the top of this tree; the *garoé* drops tiny beads of water that the people steer into a little fountain, thanks to which they and animals live through periods of extreme drought."

In 1563, Antonio Pigafetta, author of *Magellan's Voyage around the World*, put things as follows: "There is not a drop of water to be found on El Hierro, but at midday, there descends a cloud from the sky which envelops a large tree that distills a great abundance of water from its leaves and branches."

An engraving from 1748 (see p. 106) shows the Guanches, the people who inhabited the island at the time, gathering to take a shower under this tree, which they called the "fountain tree," or *garoé* in their language. The picture shows how they have set up a system for collecting water, with ditches running to their village. The tree was probably several centuries old when a storm uprooted it in 1610. The Guanches devoted a special cult to this tree, which managed to capture water from the mist and allowed them to practice agriculture despite negligible rainfall. When the *garoé* vanished, so did their culture.

Today, El Hierro has reclaimed its totem. In 1948, the thread of history was renewed when a great drought occurred: Don Zósimo Hernández Martin (1920–2004), head forest ranger on the island and an adept of both ecology and popular tradition, planted a new tree where the old one had stood. Once again, water from the fog could be gathered in

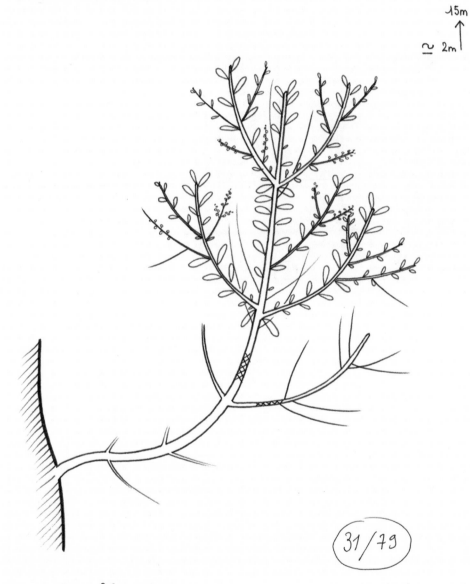

Ocotea fœtens (Spreng.) Baill.   LAURACEAE

= Oreodaphne fœtens Nees

15m

≃ 2m

31/79

~~19 V 77~~ 26 mai 1977

Villa Thuret - Iles Canaries

pits that had been dug before Spanish colonization.

In the 1960s, Don Zósimo arranged for fountain trees to be planted along the island's route of pilgrimage and cattle troughs to be built under their foliage. His project of reforestation with indigenous plants is one reason El Hierro, "The Island of Iron," was declared a UNESCO biosphere reserve in 2000.

The *garoé* isn't the only fountain tree on El Hierro; there are laurels, junipers, pines, and more. In the Sultanate of Oman, olive trees perform the same function; in northern Chile, *Tara spinosa* (Molina) Britton & Rose does the same. Catching water isn't a matter of species, but of location: it occurs where trade winds drive fog from the sea.

The *garoé* of El Hierro has inspired modern efforts to derive water from mist. In 1992, the scientific journal *La Recherche* published an article on the tree,[6] and since then, development engineers have gradually become interested in the model it offers for collecting water. In the early 2000s, nets suspended on iron structures were set up in

6. Alain Gioda, "L'arbre fontaine," *La Recherche* (December 1992): 1400–1408.

Chile to capture the fog. These artificial traps have also been used for several years now in the Canary Islands: tiny drops of water accumulate on plastic or metal netting, then drop into gutters (supplying small communities on El Hierro and the northern part of the neighboring island of Tenerife). The average yield is some thirty liters of water per square meter of netting. In the meanwhile, El Hierro has added the fountain tree, *garoé*, to its coat of arms.

"A little cloud always hangs at the top of this tree; the garoé drops tiny beads of water that the people steer into a little fountain, thanks to which they and animals live through periods of extreme drought."

Bartolomé de Las Casas, *History of the Indies*, 1559.

A CLONAL FOREST

Cyathea manniana Hook.
CYATHEACEAE
The Great African Tree-Fern

TREE FERNS are among the most beautiful plants in Africa. They're found on low-altitude mountains, at around one thousand meters. These magnificent ferns, which resemble palms, stretch as far as the eye can see and look like they're part of a vast forest. That's partly true. The base of each trunk is invisible, hidden under a thick mantle of fine, intertwining roots like felt. A cut here will reveal the actual trunk, which can be recognized by its thick diameter, and, all around it, a host of small stolons buried in the roots. After a moment's reflection, one realizes that the whole forest is one big clone; it just takes a little work picking apart the tangle of roots to see as much.

The ferns' trunks are attached to each other by stolons: stems with a small diameter and scaled-down leaves, which serve the purpose of vegetative multiplication. Hidden under the roots, they also play a stabilizing role. Each stolon grows out of the trunk and heads downward; once it gets to the ground, it steers a subterranean course for four or five meters, pops up again, and becomes the trunk of a new fern.

In addition to this process of vegetative reproduction through stolons, tree ferns also have a means of sexual reproduction: microscopic spores that enable the species to spread over great distances via the wind. Sexuality is what takes *Cyathea manniana* from one mountain to another, but stolons are responsible for the forest. These plants have existed in tropical regions since the Carboniferous Period, 345 million years before our own time. In other words, they were around long before the dinosaurs, which are only 180 million years old! The great reptiles are now gone, but tree ferns are still around—and they still look like palms.

Tropical botany may be likened to archaeology, for it often involves encountering plants that are living fossils. In part, it's the practice of present-day paleobotany; but it also concerns modern plants still evolving.

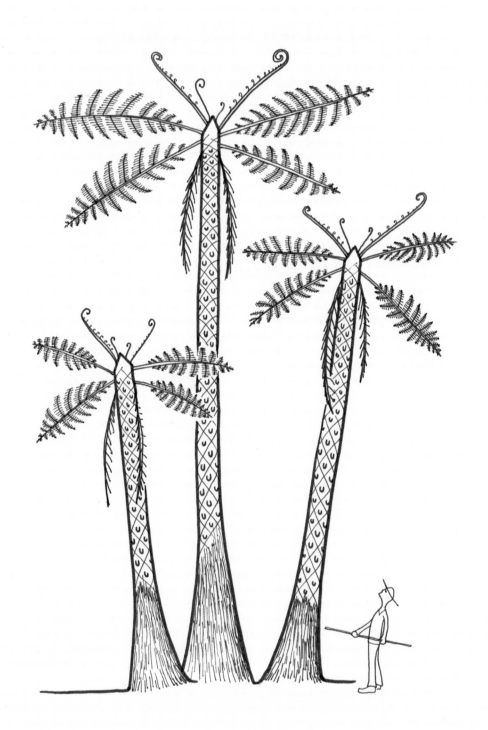

FLOWERS ON LEAVES

....

Phyllobotryon spathulatum Müll. Arg.
ACHARIACEAE
—

....

MY FATHER was an agronomist, and he traveled all over the world. His seven sons all followed in his footsteps, more or less. I'm the youngest. I remember seeing my brothers come back from the tropics with incredible tales to tell, spending a little time at home in Paris, then heading off again to fascinating, faraway lands. The tropics are a family tradition. If you're a botanist, it makes good sense to go where you'll find the most plants! Those that grow in the intertropical convergence zone are more profuse and more majestic than anywhere else. The favorable climate of forests here encourages growth and yields incomparable species diversity: thousands of different kinds of tree can be found, whereas Europe has only a hundred. Plants here are more beautiful, bigger, and more unbelievable than elsewhere—and many of them still await discovery. The plants of Europe have been studied over and over, but in forests close to the equator there are plenty without a name, which haven't even been collected yet.

Phyllobotryon is an example of the surprises the tropics hold in store. I encountered it for the first time in Campo-Ma'an National Park, Cameroon. It belongs to the family of *Achariaceae* and blossoms in surprising ways. Sometimes, the flowers shoot up from the earth (see page 42), but they can also sprout from the plant's leaves. *Phyllobotryon*'s flowers and fruit grow from the central vein on the upper side of leaves.

Normally, flowers and leaves are separate organs, as is the case with our plants in Europe. But we're talking about a tropical plant here. There are only two or three plants in the world with flowers on their leaves; *Phyllobotryon* is the prettiest.

All this makes it easy to see that European flora represent just one possibility among others. To get a sense of just how many other possibilities exist, one need only travel to a tropical rainforest.

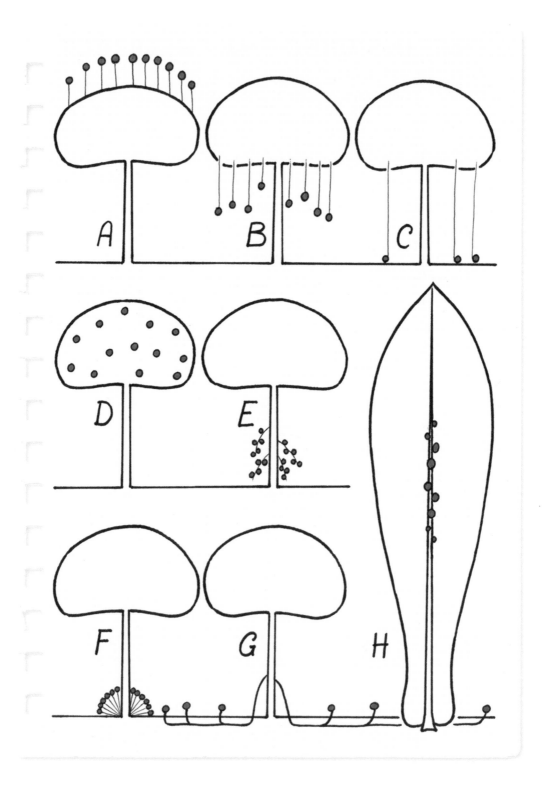

THE GIANT LILY
THAT INSPIRED THE ARCHITECT

....

Victoria amazonica (Poeppig) Sowerby
NYMPHAEACEAE
—

....

THE ONLY VISIBLE PART of this superb water lily is its leaves floating on the water. Round and huge, they can grow up to two meters in diameter. They're also flat and crenelated, like molds for giant cakes fitted out with two "overflows": little vents for the rainwater to run out.

Victoria inspired Joseph Paxton's design for the Crystal Palace. It's an interesting story. Paxton was an exceptionally talented gardener, and his whole life was intimately tied to this plant, discovered in 1801 in Bolivia. Some of its seeds made it to England in 1849, where efforts were made to grow it at the Royal Botanic Gardens, Kew, but these attempts failed. Famous for his green thumb, Paxton was called in to help, and he managed to make the last seeds sprout and successfully acclimatized the plant. When it bloomed, the event made history, and Queen Victoria made him a knight.

Paxton determined that the leaves could bear the weight of a child. The *Illustrated London News* published an engraving that showed his daughter Annie, dressed as a fairy, sitting on a specimen of *Victoria*. The plant's leaves are sturdy thanks to veins on the underside, which are organized as a system of buttresses. In turn, the radial veins, which are stiff and covered with robust thorns, are reinforced by a series of concentric, pliant fibers running in the opposite direction; the whole underside displays a magnificent purple color. Acclaimed for his gardening, Paxton was appointed architect and given the task of constructing the Crystal Palace to house the Great Exhibition. He drew inspiration from

Paxton determined that the leaves could bear the weight of a child. The Illustrated London News *published an engraving that showed his daughter Annie, dressed as a fairy, sitting on a specimen of Victoria.*

....

the leaf architecture of *Victoria* to build a magnificent castle of iron and glass, which opened in London in 1851.

Victoria amazonica offers other surprises, too. Its enormous flowers, which can measure up to thirty centimeters across, blossom for one day and two nights. The first evening, at dusk, a white flower emerges from a large bud covered with spines. It displays some hundred round petals and emits a distinctive fragrance reminiscent of pineapple. Curiously, the flower is hot: a thermochemical reaction elevates its internal temperature eleven degrees above that of its surroundings. The combination of scent and heat attracts Coleoptera; at dawn, when the flower closes again, the beetles find themselves trapped, and they spend the day feeding on starch inside the flower. During the second night, the flower becomes pink, red even. At sunset, the trapped insects regain their freedom; covered with pollen, they fly off to fertilize another flower. The next day, at dawn, the flower wilts, closes up, and vanishes; the fruit ripens underwater.

This remarkable species, a symbol of the Amazon, remains something of a mystery. We still don't know whether its flowers are terminal or lateral. What's more, its architecture as a whole requires further study—which can only be carried out by an experienced, and intrepid, diver.

THE STRANGLER FIG

Ficus
MORACEAE
Banyan

THE TROPICS are home to eight hundred fifty species of fig trees. They all bear figs, but most of them aren't edible. In Europe, we have the Mediterranean fig tree, *Ficus carica*, which bears the tastiest ones. Figs aren't fruit: just cut them open and you'll see countless, tiny flowers. Another characteristic of the fig tree is its white "latex," which flows from the slightest wound; this is true of all Moraceae.

In the tropics, these plants take on various forms: bushes, creeping plants, and tall forest trees; some of them are even epiphytes—trees that grow on other trees. This is the case with the strangler fig, which attracts birds and monkeys that swallow the seeds and spread them in their excrement.

If a seed from the strangler fig falls on the ground in the shady undergrowth, it won't yield a tree. But if it falls up on top of a tree, where there's full light, it will sprout and grow very fast. Figs contain so many seeds that it's statistically certain some will land on top of large trees. The young plant sends down roots along the whole length of the support tree describing spirals in opposite directions. As soon as they meet up on the ground, the roots fuse; in a few months, they get so big that they hold the tree in a straightjacket. The host tree can't grow anymore, and it dies after a few years. Unless it's a palm; since palm trunks don't expand in diameter, the tree can keep growing straight up and coexist peacefully with the strangler.

As a rule, when support trees die, they decompose rapidly and create compost that feeds the fig tree, which assumes enormous dimensions—much bigger than its victim's. The strangler fig is a cannibal: all that's left of the vanquished tree is a gaping void in the middle of the victor's gigantic trunk.

The Botanic Garden in Calcutta is home to a famous strangler fig. This banyan, which is said to be two hundred years old, takes up 1.6 hectares; its circumference measures 412 meters, and it weighs 1,775 tons! Other *Ficus* are capable of killing trees, albeit more discreetly. Collectively, they're called "splitters"; one of them, *Ficus religiosa*, is the Tree of the Buddha. Its roots force their way through the trunk of the support tree and reach down to the ground. As they grow, they act like a woodcutter's wedge and make the trunk of the support tree burst. Even an ordinary *Ficus carica* like the ones we have in our gardens is capable of destroying a building; it sends out fine roots that have an easy time threading between bricks and fracturing walls as they grow.

To my mind, figs represent the future of plants. The history of the Earth has witnessed different stages of flora: first came the epoch of ferns, then the age of conifers, and then the era of flowering plants—which is still our own. We know that such evolution is ongoing and that one part of the flora in our world will rise to the occasion. This might very well be *Ficus*. These trees are singular indeed, for they make fruit without displaying any flowers—the flowers are inside the figs, producing seeds that crack when eaten, the actual fruit of the tree.

Why do I venture this hypothesis? More and more, scientists are paying attention to the close coevolution that occurs between flowering plants and the insects that pollinate them. Fig trees represent a particularly advanced case: the pollinating insects live inside the figs. For each species of *Ficus*, there is just one species of insect capable of pollinating the flowers: a tiny wasp that climbs into the fig through a minuscule opening on top, then goes and deposits the pollen in another fig. There's even one kind, the caprifig ("goat fig"), which produces inedible "fruit" during the winter, whose sole purpose is to keep the pollinating wasps alive. The whole history of plants depends on ties with pollinators. Fig trees seem to stand at the forefront of this evolutionary trend.

LEXICON

ARCHITECTURAL MODEL · The architectural form of a tree isn't a matter of chance. There are twenty-four models for grouping trees together, which represent as many programs of development defined by the following criteria: whether branches grow vertically or horizontally, the stalks' mode of growth, and patterns of inflorescence. These structures are a constant for each species, however much environments may vary.

CANOPY · Vegetal layer formed by the high branches and foliage of the upper part of trees in a forest. In the tropics, canopies harbor a great number of species, plant and animal, that benefit from direct light, more or less absent at ground level.

CLONE · Collection of cells from a single original cell; collection of plants resulting from asexual reproduction.

COEVOLUTION · Evolutionary adaptation between two or more species based on reciprocal interaction (symbiosis, predator–prey, host–parasite).

EPIPHYTE · Plant that grows on another plant that serves only as support. An epiphytic plant is not a parasite because it draws its sustenance from light, air, rainwater, and debris that accumulates on the support plant.

FAMILY · Unit of classification hierarchically located above species and genus. "Family" groups together related genera. The Latin names of families end in *-aceae*.

LIANA · A rambling plant that develops a long, flexible stalk that clings to a supporting tree, either by wrapping around it or by fastening itself by means of adhesive roots, hooks, or spines.

PHOTOSYNTHESIS · Process through which chlorophyllous plants, when exposed to sunlight, synthesize organic matter from mineral substances (atmospheric carbon dioxide, minerals from the soil, water) and emit oxygen.

POLLEN · The pollen grain contains the male gametophyte that enables male gametes to form.

POLLINATION · Transfer of pollen from the male reproductive organ to the female reproductive organ of seed-bearing plants. The process generally occurs through the mediation of pollinating agents such as insects, wind, or water.

REITERATION · Mechanism by which the tree creates structures—architectural units—that it repeats at will and adds to previous units.

ROOT · Lower part of the axis of a vascular plant that serves to stabilize it and carry water and minerals. Generally subterranean, the root is distinct from the stalk inasmuch as it lacks chlorophyll and buds.

SPATHE · Membranous part of an inflorescence resembling a leaf. It is a large bract, with an opening slit, enclosing certain flowers (the iris, for instance) or inflorescences (e.g., garlic, arum).

STOLON · A shoot, often with scaled-down leaves, that grows horizontally at ground level and serves the purpose of vegetative reproduction (e.g., strawberry, potato, potentilla).

SYMBIOSIS · Stable interaction that is mutually beneficial between two partners: plants and bacteria, fungi and algae, plants and ants, plants and fungi.

TRUNK · Main stem of a tree or shrub above the root system, bearing the plant's foliage.

UNITARY · A unitary tree comprises a single architectural unit that grows larger for the duration of the plant's life while preserving its shape.